D0935623

THE ETHICAL ALGORITHM

MICHAEL KEARNS
AND
AARON ROTH

THE ETHICAL ALGORITHM

The Science of Socially Aware Algorithm Design

OXFORD
UNIVERSITY PRESS

OXFORD
UNIVERSITY PRESS

Oxford University Press is a department of the University of Oxford.
It furthers the University's objective of excellence in research, scholarship,
and education by publishing worldwide. Oxford is a registered trade mark of
Oxford University Press in the UK and certain other countries.

Published in the United States of America by Oxford University Press
198 Madison Avenue, New York, NY 10016, United States of America.

Library of Congress Cataloging-in-Publication Data
Names: Kearns, Michael, 1971– author. | Roth, Aaron (Writer on technology), author.
Title: The ethical algorithm : the science of socially aware algorithm design /
Michael Kearns and Aaron Roth.
Description: New York : Oxford University Press, 2019. |
Includes bibliographical references and index. |
Identifiers: LCCN 2019025725 |
ISBN 9780190948207 (hardback) | ISBN 9780190948221 (epub)
Subjects: LCSH: Information technology—Economic aspects. |
Technological innovations—Moral and ethical aspects.
Classification: LCC HC79.I55 K43 2019 | DDC 174/.90051—dc23
LC record available at https://lccn.loc.gov/2019025725

1 3 5 7 9 8 6 4 2
Printed by Sheridan Books, Inc.
United States of America

Dedicated to our families

MK: Kim, Kate, and Gray

AR: Cathy, Eli, and Zelda

CONTENTS

THE ETHICAL ALGORITHM

INTRODUCTION

ALGORITHMIC ANXIETY

We are allegedly living in a golden age of data. For practically any question about people or society that you might be curious about, there are colossal datasets that can be mined and analyzed to provide answers with statistical certainty. How do the behaviors of your friends influence what you watch on TV, or how you vote? These questions can be answered with Facebook data, which records the social network activity of billions of people worldwide. Are people who exercise frequently less likely to habitually check their email? For anyone who uses an Apple Watch, or an Android phone together with the Google Fit app, the data can tell us. And if you are a retailer who wants to better target your products to your customers by knowing where and how they spend their days and nights, there are dozens of companies vying to sell you this data.

Which all brings us to a conundrum. The insights we can get from this unprecedented access to data can be a great thing: we can get new understanding about how our society works, and improve public health, municipal services, and consumer products. But as individuals, we aren't just the recipients of the fruits of this data analysis: we *are* the data, and it is being used to make decisions *about us*—sometimes very consequential decisions.

In December 2018, the *New York Times* obtained a commercial dataset containing location information collected from phone apps whose nominal purpose is to provide mundane things like weather reports and restaurant recommendations. Such datasets contain precise locations for hundreds of millions of individuals, each updated hundreds (or even thousands) of times a day. Commercial buyers of such data will generally be interested in aggregate information—for example, a hedge fund might be interested in tracking the number of people who shop at a particular chain of retail outlets in order to predict quarterly revenues. But the data is recorded by individual phones. It is superficially anonymous, without names attached—but there is only so much anonymity you can promise when recording a person's every move.

For example, from this data the *New York Times* was able to identify a forty-six-year-old math teacher named Lisa Magrin. She was the only person who made the daily commute from her home in upstate New York to the middle school where she works, fourteen miles away. And once someone's identity is uncovered in this way, it's possible to learn a lot more about them. The *Times* followed Lisa's data trail to Weight Watchers, to a dermatologist's office, and to her ex-boyfriend's home. She found this disturbing and told the *Times* why: "It's the thought of people finding out those intimate details that you don't want people to know." Just a couple of decades ago, this level of intrusive surveillance would have required a private investigator or a government agency; now it is simply the by-product of widely available commercial datasets.

Clearly, we have entered a brave new world.

And it's not only privacy that has become a concern as data gathering and analysis proliferate. Because algorithms—those little bits of machine code that increasingly mediate our behavior via our phones and the Internet—aren't simply analyzing the data that we generate with our every move. They are also being used to actively make decisions that affect our lives. When you apply for a credit card, your application may never be examined by a human being. Instead, an algorithm pulling in data about you (and perhaps also about people "like you") from many different sources might automatically approve or deny your request. Though there are benefits to knowing instantaneously whether your request is approved, rather than waiting five to ten business days, this should give us a moment of pause. In many states, algorithms based on what is called machine learning are also used to inform bail, parole, and criminal sentencing decisions. Algorithms are used to deploy police officers across cities. They are being used to make decisions in all sorts of domains that have direct and real impact on people's lives. All this raises questions not only of privacy but also of fairness, as well as a variety of other basic social values including safety, transparency, accountability, and even morality.

So if we are going to continue to generate and use huge datasets to automate important decisions (a trend whose reversal seems about as plausible as our returning to an agrarian society), we have to think seriously about some weighty topics. These include limits on the use of data and algorithms, and the corresponding laws, regulations, and organizations that would determine and enforce those limits. But we must also think seriously about addressing the concerns scientifically—about what it might mean to encode ethical principles directly into the design of the algorithms that are increasingly woven into our daily lives. This book is about the emerging science of ethical algorithm design, which tries to do exactly that.

SORTING THROUGH ALGORITHMS

But first, what is an algorithm anyway? At its most fundamental level, an algorithm is nothing more than a very precisely specified series of instructions for performing some concrete task. The simplest algorithms—the ones we teach to our first-year computer science students—do very basic but often important things, such as sorting a list of numbers from smallest to largest. Imagine you are confronted with a row of a billion notecards, each of which has an arbitrary number written on it. Your goal is to rearrange the notecards so that the numbers are in ascending order—or, more precisely, to specify an algorithm for doing so. This means that each step of the process you describe must be unambiguous, and that the process must always terminate with the notecards arranged in ascending order, regardless of the numbers and their initial arrangement.

Here is one way to start. Scan through the initial arrangement from left to right to find the smallest of the numbers (you're allowed to use a pencil and paper as storage or memory to help you), which perhaps is written on the 65,704th notecard. Then swap that notecard with the leftmost notecard. Now the smallest number comes first in the list, as desired. Next, rescan the cards starting second from left to find the second-smallest number, and swap its notecard with the second from left. Continue in this fashion until you've completely sorted the list. This is an algorithm—it's precisely specified, and it always works.

It's also a "bad" algorithm, in the sense that there are considerably faster algorithms for the same problem. If we think about it for a moment, we see that this algorithm will scan through the list of a billion numbers a billion times—first from leftmost to rightmost, then from second from left to rightmost, then from third from left to rightmost, and so on—each time placing just one more number in its proper position. In the language of algorithms, this would be called a quadratic time algorithm, because if the length of the list is n, the number of steps or "running time" required by the algorithm would be proportional to the square of n. And if n is in the billions—as it would be, for example, if we wanted to sort Facebook users by their monthly usage time—it would be infeasibly slow for even the fastest computers. Fortunately for Facebook (and the rest of us), there are algorithms whose running time is much closer to n than to its square. Such algorithms are fast enough for even the largest real-world sorting problems.

One of the interesting aspects of algorithm design is that even for fundamental problems such as sorting, there can be multiple alternative algorithms with different strengths and weaknesses, depending on what our concerns are. For example, the extensive Wikipedia page on sorting lists forty-three different algorithms, with names like Quicksort, Heapsort, Bubblesort, and Pigeonhole Sort. Some are faster when we can assume the initial list is in a random order (as opposed to being in reverse sorted order, for example). Some require less memory than others, at the expense of being slower. Some excel when we can assume that each number in the list is unique (as with social security numbers).

So even within the constraint of developing a precise recipe for a precise computational task, there may be choices and trade-offs to confront. As the previous paragraph suggests, computer science has traditionally focused on algorithmic trade-offs related to what we might consider performance metrics, including computational speed, the amount of memory required, or the amount of communication required between algorithms running on separate computers. But this book, and the emerging research it describes, is about an entirely new dimension in algorithm design: the explicit consideration of social values such as privacy and fairness.

MAN VERSUS MACHINE (LEARNING)

Algorithms such as the sorting algorithm we describe above are typically coded by the scientists and engineers who design them: every step of the procedure is explicitly specified by its human designers, and written down in a general-purpose programming language such as Python or C++. But not all algorithms are like this. More complicated algorithms—the type that we categorize as machine learning algorithms—are automatically derived from data. A human being might hand-code the process (or meta-algorithm) by which the final algorithm—sometimes called a model—is derived from the data, but she doesn't directly design the model itself.

In traditional algorithm design, while the output might be useful (like a sorted list of Facebook usage times, which could help in analyzing the demographic properties of the most engaged users), that output is not itself another algorithm that can be directly applied to further data. In contrast, in machine learning, that's the entire point. For example, think about taking a database of high school information about previously admitted college students, some of whom graduated from college and some of whom did not, and using it to derive a model predicting the likelihood of graduation for future applicants. Rather than trying to directly specify an algorithm for making these predictions—which could be quite difficult and subtle—we write a meta-algorithm that uses the historical data to *derive* our model or prediction algorithm. Machine learning is sometimes considered a form of "self-programming," since it's primarily the data that determines the detailed form of the learned model.

This data-driven process is how we get algorithms for more human-like tasks, such as face recognition, language translation, and lots of other prediction problems that we'll talk about in this book. Indeed, with the aforementioned explosion of consumer data enabled by the Internet, the machine learning approach to algorithm design is now much more the rule than the exception. But the less directly involved humans are with the final algorithm or model, the less aware they may be of the unintended ethical, moral, and other side effects of those models, which are the focus of this book.

HOW THINGS CAN GO WRONG

The reader might be excused for some skepticism about imparting moral character to an algorithm. An algorithm, after all, is just a human artifact, like a hammer, and who would entertain the idea of an ethical hammer? Of course, a hammer might be put to an unethical use—as an instrument of violence, for example—but this can't be said to be the hammer's fault. Anything ethical about the use or misuse of a hammer can be attributed to the human being who wields it.

But algorithms—especially models derived directly from data via machine learning—are different. They are different both because we allow them a significant amount of agency to make decisions without human intervention and because they are often so complex and opaque that even their designers cannot anticipate how they will behave in many situations. Unlike a hammer, which is designed to do only one thing exceptionally well, algorithms can be tremendously general-purpose, closer to the human mind in their flexibility of purpose than to something you'd find in a carpenter's toolbox. And unlike with a hammer, it is usually not so easy to blame a particular misdeed of an algorithm directly on the person who designed or deployed it. In this book, we will see many instances in which algorithms leak sensitive personal information or discriminate against one demographic or another. But how exactly do these things happen? Are violations of privacy and fairness the result of incompetent software developers or, worse yet, the work of evil programmers deliberately coding racism and back doors into their programs?

The answer is a resounding no. The real reasons for algorithmic misbehavior are perhaps even more disturbing than human incompetence or malfeasance (which we are at least more familiar with and have some mechanisms for addressing). Society's most influential algorithms—from Google search and Facebook's News Feed to credit scoring and health risk assessment algorithms—are generally developed by highly trained scientists and engineers who are carefully applying well-understood design principles. The problems actually lie within those very principles, most specifically those of machine learning. It will serve us well later to discuss those principles a bit now.

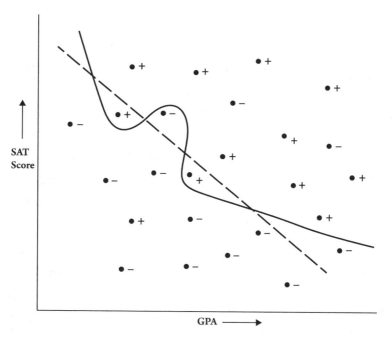

Fig. 1. Building a model to predict collegiate success from high school data. Imagine that each point represents the high school GPA and SAT score of a college student. Points labeled with "+" represent students who successfully graduated from college in four years, while points labeled with "–" represent students who did not. The straight dashed line does an imperfect but pretty good job of separating positives from negatives and could be used to predict success for future high school students. The solid curve makes even fewer errors but is more complicated, potentially leading to unintended side effects.

As we've suggested, many of the algorithms we discuss in this book would more accurately be called *models*. These models, which make the actual decisions of interest, are the result of powerful machine learning (meta-) algorithms being applied to large, complex datasets. A crude but useful sketch of the pipeline is that the data is fed to an algorithm, which then searches a very large space of models for one that provides a good fit to the data. Think of being given a cloud of 100 points on a piece of paper, each labeled either "positive" or "negative," and being asked to draw a curve that does a good but perhaps imperfect job of separating positives from negatives (see Figure 1). The positive and negative points are the data, and you are the algorithm— trying out different curves until you settle on what you think is the best separator. The curve you pick is the model, and it will be used to predict whether future points are positive or negative But now imagine that instead of 100 points, there are 10 million; and instead of the points being on a 2-dimensional sheet of paper, they lie in a 10,000-dimensional space. We can't expect you to act as the algorithm anymore, however smart you might be.

The standard and most widely used meta-algorithms in machine learning are simple, transparent, and principled. In Figure 2 we replicate the high-level description or "pseudocode" from Wikipedia for the famous backpropagation algorithm for neural networks, a powerful class of predictive models. This description is all of eleven lines long, and it is easily taught to undergraduates. The main "forEach" loop is simply repeatedly cycling through the data points (the positive and negative dots on the page) and adjusting the parameters of the model (the curve you were fitting) in an attempt to reduce the number of misclassifications (positive points the model misclassifies as negative, and negative points the model misclassifies as positive). It's doing what you would do, except in a perhaps more systematic, mathematical way, and without any limits on how many data points there are or how complex they are.

```
initialize network weights (often small random values)
do
    forEach training example named ex
        prediction = neural-net-output(network, ex)   // forward pass
        actual = teacher-output(ex)
        compute error (prediction - actual) at the output units
        compute Δw_h for all weights from hidden layer to output layer   // backward pass
        compute Δw_i for all weights from input layer to hidden layer     // backward pass continued
        update network weights // input layer not modified by error estimate
until all examples classified correctly or another stopping criterion satisfied
return the network
```

Fig. 2. Pseudocode for the backpropagation algorithm for neural networks.

So when people talk about the complexity and opaqueness of machine learning, they really don't (or at least shouldn't) mean the actual optimization algorithms, such as backpropagation. These are the algorithms designed by human beings. But the *models* they produce— the outputs of such algorithms—can be complicated and inscrutable, especially when the input data is itself complex and the space of possible models is immense. And this is why the human being deploying the model won't fully understand it. The goal of backpropagation is perfectly understandable: minimize the error on the input data. The opacity of machine learning, and the problems that can arise, are really emergent phenomena that result when straightforward algorithms are allowed to interact with complex data to produce complex predictive models.

For example, it may be that the model that minimizes the overall error in predicting collegiate success, when used to make admissions decisions, happens to falsely reject qualified black applicants more often than qualified white applicants. Why? Because the designer didn't anticipate it. She didn't tell the algorithm to try to equalize the false rejection rates between the two groups, so it didn't. In its standard form, machine learning won't give you anything "for free" you didn't explicitly ask for, and may in fact often give you the opposite of what you wanted. Put another way, the problem is that rich model spaces such as neural networks may contain many "sharp corners" that provide the opportunity to achieve their objective at the expense of other things we didn't explicitly think about, such as privacy or fairness.

The result is that the complicated, automated decision-making that can arise from machine learning has a character of its own, distinct from that of its designer. The designer may have had a good understanding of the algorithm that was used to *find* the decision-making model, but not of the model itself. To make sure that the effects of these models respect the societal norms that we want to maintain, we need to learn how to design these goals directly into our algorithms.

WHO ARE WE?

Before we embark on our journey intertwining technology, society, ethics, and algorithm design, it will be helpful to know a little about who we are and how we came to be interested in these apparently disparate topics. Our backgrounds will in turn illuminate what this book is and is not intended to be about, and what we are more and less qualified to opine upon.

For starters, we are both career researchers in the field known as theoretical computer science. As the name somewhat generically suggests, this is the branch of computer science with particular interest in formal, mathematical models of computation. We deliberately say "computation" and not "computers," because for the purposes of this book (and perhaps even generally), the most important thing to know about theoretical computer science is that it views computation as a ubiquitous phenomenon, not one that is limited to technological artifacts. The scientific justification for this view originates with the staggeringly influential work of Alan Turing (the first theoretical computer scientist) in the 1930s, who demonstrated the universality of computational principles with his mathematical model now known as the Turing machine. Many trained in theoretical computer science, ourselves included, view the field and its tools not simply as another scientific discipline but as a way of seeing and understanding the world around us—perhaps much as those trained in theoretical physics in an earlier era saw their own field.

So a theoretical computer scientist sees computation taking place everywhere—certainly in computers, but also in nature (in genetics, evolution, quantum mechanics, and neuroscience), in society (in markets and other systems of collective behavior), and beyond. These areas all embody computation in the general sense that Turing envisioned. Certainly the physical mechanisms and details differ—computation in genetics, for example, involves DNA and RNA instead of the circuits and wires of traditional electronic computers, and is less precise—but we can still extract valuable insights by treating such varied systems as computational devices.

This worldview is actually shared by many computer scientists, not only the theoretical ones. The distinguishing feature of theoretical computer science is the desire to formulate mathematically precise models of computational phenomena and to explore their algorithmic consequences. A machine learning practitioner might develop or take an algorithm like backpropagation for neural networks, which we discussed earlier, and apply it to real data to see how well it performs. Doing so doesn't really require the practitioner to precisely specify what "learning" means or doesn't mean, or what computational difficulties it might present generally. She can simply see whether the algorithm works well for the specific data or task at hand.

In contrast, a theoretical computer scientist would be inclined to approach machine learning by first giving a precise definition (or perhaps multiple variations on an underlying definition) of "learning," and then to systematically explore what can and cannot be achieved algorithmically under this definition. With tongue only slightly in cheek, we can view the typical practitioner as following Nike's "Just Do It" mantra, while the theorist follows the "Just Define and Study It" mantra. When put this way, people might naturally wonder what the practical value of theoretical computer science is, and it's true that in many scientific disciplines theory often lags behind practice. But we would argue (and suspect many of our colleagues would agree) that the theoretical approach is essential when the right defini-tion is far from clear, and when getting it right matters a lot, as with concepts such as "privacy" and "fairness."

Writing down precise definitions that capture the essence of critical and very human ideas without becoming overly complex is something of an art form, and it is inevitable that in many settings, simplifications—sometimes painful ones—are necessary. We will see this tension arise repeatedly throughout this book. But we should keep in mind that this tension is not an artifact of the theoretical approach per se; instead, it reflects the inherent difficulty of being precise about concepts that previously have been left vague, such as "fairness." We believe that the only way to make algorithms better behaved is to begin by specifying what our goals for them might be in the first place.

Our training notwithstanding, our research and interest in the topics described here have not been formulated in a vacuum of abstraction and mathematics. We've both always been interested in applying that approach to problems in machine learning and artificial intelligence. We are also neither adverse to nor inexperienced in experimental, data-driven work in machine learning—often as a test of the practicality and limitations of our theories, but not always. And it was the very trends we describe in these pages—the explosive growth of consumer data enabled by the Internet, and the accompanying rise in machine learning for automated decision-making—that made us and our col-leagues aware of and concerned about the potential collateral damage.

We have spent much of the last decade researching the topics that we cover in this book, and engaging with a variety of stakeholders. We've spent many hours talking to lawyers, regulators, economists, criminologists, social scientists, technology industry professionals, and many others about the issues raised in these pages. We've provided testimony and input to congressional committees, corporations, and government agencies on algorithmic privacy and fairness. And between us we have extensive, hands-on professional experience in areas including quantitative trading and finance; legal, regulatory, and algorithmic consulting; and technology investing and start-ups—all of which are beginning to confront the social issues that are our themes here.

We are, in short, modern computer scientists. We also know what we are not, and should not pretend to be. We are not lawyers or regulators. We are not judges, police officers, or social workers. We are not on the front lines, directly seeing and helping the people who suffer harms from privacy or fairness violations by algorithms. We are not social activists with a deep, firsthand sense of the history and problems of discrimination and other forms of injustice.

Because of this, we'll tend to say relatively little on important matters where such expertise is essential, such as designing better laws or policies, proposing how to improve social agencies to reduce unfairness in the first place, or opining on whether and how to stem labor displacement resulting from technology. It's not that we don't care about these topics or have opinions on them; we do. But doing justice to them would have fundamentally altered the nature of the book we wanted to write, which is about scientific approaches and solutions to what we might call socio-algorithmic problems.

So we will stick to what we know well and have thought hard about: how to design better algorithms.

WHAT THIS BOOK IS (NOT) ABOUT

It doesn't take much searching to discover many recent books, news stories, and scientific articles documenting cases in which algorithms have caused harms to specific people, and often to large groups of people. For instance, controlled online experiments have demonstrated racial, gender, political, and other types of bias in Google search results, Facebook advertising, and other Internet services. An explosive controversy over racial discrimination in the predictive models used in criminal sentencing decisions has recently consumed statisticians, criminologists, and legal scholars. In the domain of data privacy, there have been many cases in which sensitive information about specific people—including medical records, web search behavior, and financial data—has been inferred by "de-anonymizing" data that was allegedly made "anonymous" by algorithmic processing (as in the aforementioned *New York Times* case about location data and Lisa Magrin). Turbocharged by algorithmic data analysis tools that make it faster and more efficient to search for correlations in data, there has even been a rash of reported scientific findings that turn out not to be true, costing both dollars and lives. It has become very clear that modern algorithms may routinely trample on some of our most cherished social values.

So the problems are becoming obvious. What about the solutions? Much of the discussion to date has focused on what we might consider to be "traditional" solutions, such as new laws and regulations focused on algorithms, data and machine learning. The European Union's General Data Protection Regulation is a sweeping set of laws designed to limit algorithmic violations of privacy and to enforce still-vague social values such as "accountability" and "interpretability" on algorithmic behavior. Legal scholars are immersed in discussions of the ways existing laws do or do not apply to previously human-centric arenas that are increasingly dominated by algorithms, such as Title VII's prohibition of discrimination in employment decisions in the United States. The tech industry itself is starting to develop self-regulatory initiatives of various types, such as the Partnership on AI to Benefit People and Society. Government organizations and regulatory agencies are struggling to figure out how the algorithmic landscape affects their missions; the US State Department even held a workshop on the role and influence of AI in foreign policy.

There are important conversations going on, even as we write, about the proper role of data collection in our society: maybe there are certain things that just shouldn't be done, because the long-term social consequences aren't worth the gains. It might be beside the point whether or not facial recognition algorithms have higher error rates on black people compared to white people; maybe we shouldn't be engaging in large-scale facial recognition at all, simply because it leads us closer to being a surveillance state. All of this activity and debate is healthy, important, and essential and has been written about at length by others.

So rather than covering familiar ground, our book instead dives headfirst into the emerging science of designing social constraints directly into algorithms, and the consequences and trade-offs that emerge. Despite our focus on concrete technical solutions in the rest of this book, we are not under the misapprehension that technology alone can solve complicated social problems. But neither can decisions about our society be made in a vacuum. To make informed decisions, we need to be able to understand the consequences of deploying certain kinds of algorithms, and the costs associated with constraining them in various ways. And that is what this book is about.

At this juncture you'd be forgiven for feeling a tad queasy about a book on ethical algorithms written by theoretical computer scientists. The same people who brought you the disease have now proposed the cure—and it's more algorithms! But we indeed believe that curtailing algorithmic misbehavior will itself require more and better algorithms—algorithms that can assist regulators, watchdog groups, and other human organizations to monitor and measure the undesirable and unintended effects of machine learning and related technologies. It will also require versions of those technologies that are more "socially aware" and thus better-behaved in the first place. This book is about the new science underlying algorithms that internalize precise definitions of things such as fairness and privacy—specified by humans—and make sure they are obeyed. Instead of people regulating and monitoring algorithms from the outside, the idea is to fix them from the inside. To use one of the nascent field's acronyms, the topics we examine here are about the FATE—fairness, accuracy, transparency, and ethics—of algorithm design.

There is no getting around the fact that the people who have developed a particular branch of science or technology in the first place are almost always the ones most deeply familiar with its limitations, drawbacks, and dangers—and possibly how to correct or reduce them. So it's essential that the scientific and research communities who work on machine learning be engaged and centrally involved in the ethical debates around algorithmic decision-making. Consider the Manhattan Project to develop the atomic bomb during World War II, for instance, and the subsequent efforts of many of its scientists to curtail the use of their own invention for years after. Of course, the loss of human life from algorithms has, mercifully, been far less severe (at least so far) than from the use of nuclear weapons, but the harms are more diffuse and harder to detect. Whatever one believes about the ultimate role that algorithms should play in our society, the idea that their designers should inform that role is fundamentally sound.

The efforts described in these pages in no way propose that algorithms themselves should decide, or even be used to decide, the social values they will enforce or monitor. Definitions of fairness, privacy, transparency, interpretability, and morality should remain firmly in the human domain—which is one of the reasons that the endeavor we describe must ultimately be a collaboration between scientists, engineers, lawyers, regulators, philosophers, social workers, and concerned citizens. But once a social norm such as privacy is given a precise, quantitative definition, it can be "explained" to an algorithm that then makes sure to obey it.

Of course, one of the greatest challenges here is in the development of quantitative definitions of social values that many of us can agree on. And we'll see that this challenge has been met relatively well so far (if inevitably imperfectly) in areas such as privacy, is making good but murkier progress in areas such as fairness, and has a much longer way to go for values such as interpretability or morality. But despite the difficulties, we argue that the effort to be exceedingly precise by what we mean when we use words such as *privacy* and *fairness* has great merit in its own right—both because it is necessary in the algorithmic era and because doing so often reveals hidden subtleties, flaws, and trade-offs in our own intuitions about these concepts.

A BRIEF PREVIEW

In this book we will see how it is possible to expand the principles on which machine learning is based to demand that they incorporate—in a quantitative, measurable, verifiable manner—many of the ethical values we care about as individuals and as a society.

Of course, the first challenge in asking an algorithm to be fair or private is agreeing on what those words should mean in the first place—and not in the way a lawyer or philosopher might describe them, but in so precise a manner that they can be "explained" to a machine. This will turn out to be both nontrivial and revealing, since many of the first definitions we might consider turn out to have serious flaws. In other cases we will see there may be several intuitive definitions that are actually in conflict with each other.

But once we've settled on our definitions, we can try to internalize them in the machine learning pipeline, encoding them into our algorithms. But how? Machine learning already has a "goal," which is to maximize predictive accuracy. How do we introduce new goals such as fairness and privacy into the code without "confusing" the algorithm? The short answer is that we view these new goals as constraints on the learning process. Instead of asking for the model that only minimizes error, as we ask for the model that minimizes error *subject to the constraint* that it not violate particular notions of fairness or privacy "too much." While this might be a harder problem computationally, conceptually it is only a slight variation on the original—but a variation with major consequences.

The first major consequence is that we will now have algorithms that are guaranteed to have the particular ethical behaviors we asked for. But the second major consequence is that these guarantees will come at a cost—namely, a cost in the accuracy of the models we learn. If the most accurate model for predicting loan repayment is racially biased, then, by definition, eradicating that bias results in a less accurate model. Such costs may present hard decisions for companies, their regulators, their users, and society at large. How will we feel if more fair and private machine learning results in worse search results from Google, less efficient traffic navigation from Waze, or worse product recommendations from Amazon? What if asking for fairness from criminal sentencing models means more criminals are freed or a greater number of innocent people are incarcerated?

The good news is that the trade-offs between accuracy and good behavior can also be quantified, allowing stakeholders to make informed decisions. In particular, we will see that both accuracy and the social values we consider can be put on sliding scales that are under our control. We'll see concrete examples of this when it comes to privacy in Chapter 1 and fairness in Chapter 2.

Our discussion so far largely refers to the use of machine learning in an isolated, sequential fashion, in which models are built from personal data in order to make predictions or decisions about future individuals. But there are many settings in which there are complex feedback loops between users, the data they generate, the models constructed, and the behavior of those users. Navigation apps use our GPS data to model and predict traffic, which in turn influences the driving routes they suggest to us, which then alters the data used to build the next model. Facebook's News Feed algorithm uses our feedback to build models of the content we want to see, which in turn influences the articles and posts we read and "like," which again changes the model. The entire system of users, data, and models is perpetually changing and evolving, often in self-interested and strategic ways that we'll examine. We'll even take this view of scientific research itself, as well as its recent "reproducibility crisis." In order to understand such systems, and design them in ways that encourage good social outcomes, we will require some additional science that marries algorithm design with economics and game theory.

What we've just outlined already provides a crude road map to the rest of the book. Chapters 1 and 2 will consider algorithmic privacy and fairness in turn. In our view, these are the areas of ethical algorithms in which the most is known, and where there are relatively mature frameworks and results to discuss. Chapter 3 considers strategic feedback loops between users, data and, algorithms, but it is connected to the previous chapters by its focus on the societal consequences of algorithmic behavior. This leads us to Chapter 4, which focuses on data-driven scientific discovery and its modern pitfalls. Chapter 5 provides some brief thoughts on ethical issues that we consider important but about which we feel there is less to say scientifically so far—issues such as transparency, accountability, and even algorithmic morality. Finally, in the conclusion we briefly distill some lessons learned.

It is important to emphasize at the outset that it will never alone suffice to formalize the social values we care about and design them into our algorithms—it is also essential that such algorithms actually be adopted on a large scale. If platform companies, app developers, and government agencies don't care about privacy or fairness (or if in fact those norms run counter to their objectives), then without encouragement, pressure or coercion they will ignore the types of algorithms we will be describing in these pages. And in our current technological environment, it may often feel like there is indeed a great mismatch between the values of society and the corporations and organizations that collect and control our data. But this is not a reason not to do and understand the science now, and many recent developments—including widespread calls for data and algorithmic regulation, increasing pressure from consumers and legislature around anti-social algorithmic behavior, and greater layperson awareness of the harms being inflicted—suggest that the need for the science may arrive sooner rather than later.

The science of ethical algorithms is in its infancy; most of the research we describe in this book is less than a decade old, and some of it is far younger. And it is something in which we have immersed ourselves over many years, helping to advance the science described in this book. These are fast-moving fields. It's a certainty that much of what we cover will be outdated and updated in the near future. To some that might be a sign that it's too early to write an overview of the field; to us it's the opposite, because the most exciting times in science and technology are when the ground beneath you is shifting rapidly. We shall grapple here with difficult and consequential issues, but at the same time we want to communicate the excitement of new science. It is in this spirit of uncertainty and adventure that we begin our investigations.

1

Algorithmic Privacy

From Anonymity to Noise

"ANONYMIZED DATA ISN'T"

It has been difficult for medical research to reap the fruits of large-scale data science because the relevant data is often highly sensitive individual patient records, which cannot be freely shared. In the mid-1990s, a government agency in Massachusetts called the Group Insurance Commission (GIC) decided to help academic researchers by releasing data summarizing hospital visits for every state employee. To keep the records anonymous, the GIC removed explicit patient identifiers, such as names, addresses, and social security numbers. But each hospital record did include the patient's zip code, birthdate, and sex—these were viewed as useful summary statistics that were coarse enough that they could not be mapped to specific individuals. William Weld, the governor of Massachusetts, assured voters that the removal of explicit patient identifiers would protect patient privacy.

Latanya Sweeney, who was a PhD student at MIT at the time, was skeptical. To make her point, she set out to find William Weld's medical records from the "anonymous" data release. She spent $20 to purchase the voter rolls for the city of Cambridge, Massachusetts, where she knew that the governor lived. This dataset contained (among other things) the name, address, zip code, birthdate, and sex of every Cambridge voter—including William Weld's. Once she had this information, the rest was easy. As it turned out, only six people in Cambridge shared the governor's birthday. Of these six, three were men. And of these three, only one lived in the governor's zip code. So the "anonymized" record corresponding to William Weld's combination of birthdate, sex, and zip code was unique: Sweeney had identified the governor's medical records. She sent them to his office.

In retrospect, the problem was that although birthdate, sex, and zip code could not be used individually to identify particular individuals, in combination they could. In fact, Sweeney subsequently estimated from US Census data that 87 percent of the US population can be uniquely identified from this data triple. But now that we know this, can the problem of privacy be solved by simply concealing information about birthdate, sex, and zip code in future data releases?

It turns out that lots of less obvious things can also identify you— like the movies you watch. In 2006, Netflix launched the Netflix Prize competition, a public data science competition to find the best "collaborative filtering" algorithm to power Netflix's movie recommendation engine. A key feature of Netflix's service is its ability to recommend to users movies that they might like, given how they have rated past movies. (This was especially important when Netflix was primarily a mail-order DVD rental service, rather than a streaming service— it was harder to quickly browse or sample movies.) Collaborative filtering is a kind of machine learning problem designed to recommend purchases to users based on what similar users rated well. For every user, Netflix had the list of movies that person had rated. For each movie, Netflix knew both what score the user had given the movie (on a scale of 1 to 5 stars) and the date on which the user provided the rating. The goal of a collaborative filtering engine is to predict how a given user will rate a movie she hasn't seen yet. The engine can then recommend to a user the movies that it predicts she will rate the highest.

Netflix had a basic recommendation system based on collaborative filtering, but the company wanted a better one. The Netflix Prize competition offered $1 million for improving the accuracy of Netflix's existing system by 10 percent. A 10 percent improvement is hard, so Netflix expected a multiyear competition. An improvement of 1 percent over the previous year's state of the art qualified a competitor for an annual $50,000 progress prize, which would go to the best recommendation system submitted that year. Of course, to build a recommendation system, you need data, so Netflix publicly released a lot of it—a dataset consisting of more than a hundred million movie rating records, corresponding to the ratings that roughly half a million users gave to a total of nearly eighteen thousand movies.

Netflix was cognizant of privacy concerns: it turns out that in the United States, movie rental records are subject to surprisingly tough privacy laws. The Video Privacy Protection Act was passed by the United States Congress in 1988, after Robert Bork's video rentals were published in the *Washington City Paper* during his Supreme Court nomination hearings. The law holds any video rental provider liable for up to $2,500 in damages per customer whose records are released. So, just as the state of Massachusetts had done, Netflix removed all user identifiers. Each user was instead represented with a unique but meaningless numeric ID. This time there was no demographic information at all: no gender, no zip code. The only data about each user was his or her movie ratings.

Yet only two weeks after the data release, Arvind Narayanan (who was a PhD student at the University of Texas at Austin) and his advisor, Vitaly Shmatikov, announced that they could attach real names to many of the "anonymized" Netflix records. In their research paper, they wrote:

> We demonstrate that an adversary who knows only a little bit about an individual subscriber can easily identify his or her record if it is present in the dataset, or, at the very least, identify a small set of records which include the subscriber's record. The adversary's background knowledge need not be precise, e.g., the dates may only be known to the adversary with a 14-day error, the ratings may be known only approximately, and some of the ratings and dates may even be completely wrong.

Regarding the Netflix dataset, they found that if an attacker knew the approximate dates (give or take a couple of weeks) when a target had rated six movies, he could uniquely identify that person 99 percent of the time. They also showed that this could be done at a large scale, by cross-referencing the Netflix dataset with Internet Movie Database (IMDB) movie ratings—which people post publicly under their own names.

But if people are posting about which movies they watch publicly, can identifying them in the Netflix dataset really be considered a privacy violation? Yes. People may publicly review only a small fraction of the movies they watch, but rate all of them on Netflix. The public reviews may suffice to reveal their Netflix identity, which then unveils all the movies they have watched and rated—which can expose sensitive information, including political leanings and sexual orientation. In fact, a gay mother of two who was not open about her sexual orientation sued Netflix, alleging that the ability to de-anonymize the dataset "would negatively affect her ability to pursue her livelihood and support her family, and would hinder her and her children's ability to live peaceful lives." She was worried that her sexual orientation would become clear if people knew what movies she watched on Netflix. The lawsuit sought the maximum penalty allowed by the Video Privacy Protection Act: $2,500 for each of Netflix's more than two million subscribers. Netflix settled this lawsuit for undisclosed financial terms, and cancelled a planned second Netflix Prize competition.

The history of data anonymization is littered with many more such failures. The problem is that a surprisingly small number of apparently idiosyncratic facts about you—like when you watched a particular movie, or the last handful of items you purchased on Amazon—are enough to uniquely identify you among the billions of people in the world, or at least among those appearing in a large database. When a data administrator is considering releasing an "anonymized" data source, he can try to make an informed guess about how difficult it would be for an attacker to reidentify individuals in the dataset. But it can be difficult for him to anticipate other data sources—like the IMDB reviews—that might be available. And once data is released on the Internet, it cannot be revoked in any practical sense. So the data administrator must be able to anticipate not just every attack that can be launched using other data sources that are currently available but also attacks that might use data sources that become available in the future. This is essentially an impossible task, and the reason that Cynthia Dwork (one of the inventors of differential privacy, which we will discuss later in this chapter) likes to say that "anonymized data isn't"—either it isn't really anonymous or so much of it has been removed that it is no longer data.

A BAD SOLUTION

How can we fix the problem of de-anonymization, also known as re-identification? In both the Massachusetts hospital records and Netflix examples, the trouble was that there were lots of unique records in the dataset. Only one male Cambridge resident living in the 02139 zip code was born on December 18, 1951; only one Netflix user had watched *Hour of the Wolf, Brazil, Matinee*, and *The City of Lost Children* in March 2005. The problem with unique records is that they are akin to a fingerprint that can be reconnected with an identity by anyone who knows enough about someone to identify that person's record—and then she can learn everything else contained in that record.

For example, consider the following fictional table of patients at the Hospital of the University of Pennsylvania (HUP). Even with names removed, anyone who knows that Rebecca is a patient at HUP who also knows her age and gender can learn that she has HIV, because those attributes uniquely identify her within this database.

Name	Age	Gender	Zip Code	Smoker	Diagnosis
Richard	64	Male	19146	Y	Heart disease
Susan	61	Female	19118	N	Arthritis
Matthew	67	Male	19104	Y	Lung cancer
Alice	63	Female	19146	N	Crohn's disease
Thomas	69	Male	19115	Y	Lung cancer
Rebecca	*56*	*Female*	19103	N	HIV
Tony	52	Male	19146	Y	Lyme disease
Mohammed	59	Male	19130	Y	Seasonal allergies
Lisa	55	Female	19146	N	Ulcerative colitis

Hypothetical database of patient records, in which there is only one 56-year-old female, thus enabling anyone who knows Rebecca and the fact that she is a patient to infer that she has HIV.

An initial idea for a solution, called *k*-anonymity, is to redact information from individual records so that no set of characteristics matches just a single data record. Individual characteristics are divided into "sensitive" and "insensitive" attributes. In our fictional table, the diagnosis is sensitive, and everything else is insensitive. The goal of

k-anonymity is to make it hard to link insensitive attributes to sensitive attributes. Informally, a released set of records is k-anonymous if any combination of insensitive attributes appearing in the database matches at least k individuals in the released data. There are two main ways to redact information in a table to make it k-anonymous: we can suppress information entirely (that is, not include it at all in the released data), or we can coarsen it (not release the information as precisely as we know it, but instead bucket it). Consider the following redacted table of our fictional HUP patients:

Name	Age	Gender	Zip Code	Smoker	Diagnosis
*	60–70	Male	191**	Y	Heart disease
*	60–70	Female	191**	N	Arthritis
*	60–70	Male	191**	Y	Lung cancer
*	60–70	Female	191**	N	Crohn's disease
*	60–70	Male	191**	Y	Lung cancer
*	*50–60*	*Female*	191**	N	HIV
*	50–60	Male	191**	Y	Lyme disease
*	50–60	Male	191**	Y	Seasonal allergies
*	*50–60*	*Female*	191**	N	Ulcerative colitis

The same database after 2-anonymous redactions and coarsening. Now two records match Rebecca's age range and gender.

The table has been modified: names have been redacted, and ages and zip codes have been "coarsened" (ages are now reported as ten-year ranges, and zip codes are reported only to the first three digits). The result is that the table is now 2-anonymous. Any insensitive information we might know about a person—for example, that Rebecca is a fifty-six-year old female—now corresponds to at least two different records. So no person's record can be uniquely reidentified from his or her insensitive information.

Unfortunately, although k-anonymity can prevent record reidentification in the strictest sense, it highlights that reidentification is not the only (or even the main) privacy risk. For example, suppose we know that Richard is a sixty-something male smoker who is a patient at HUP. We can't identify his record—this information corresponds to three records in the modified table. But two of those records correspond to patients with lung cancer, and one corresponds to a patient with heart disease—so we can be sure that Richard has either lung cancer or heart disease. This is a serious privacy violation, and not one that is prevented by *k*-anonymity.

But the concept of *k*-anonymity also suffers from an even worse and more subtle problem, which is that its guarantees go away entirely when multiple datasets are released, even if all of them have been released with a *k*-anonymity guarantee. Suppose that in addition to knowing that Rebecca is a fifty-six-year-old female nonsmoker, we know that she has been a patient at two hospitals: HUP and the nearby Pennsylvania Hospital. And suppose that both hospitals have released *k*-anonymous patient records. HUP released the 2-anonymous records we saw above, and Pennsylvania Hospital released the following 3-anonymous table:

Name	Age	Gender	Zip Code	Diagnosis
*	*50–60*	*Female*	191**	HIV
*	*50–60*	*Female*	191**	Lupus
*	*50–60*	*Female*	191**	Hip fracture
*	60–70	Male	191**	Pancreatic cancer
*	60–70	Male	191**	Ulcerative colitis
*	60–70	Male	191**	Flu-like symptoms

In this 3-anonymous database from a different hospital in which Rebecca is a patient, three records match her age and gender. By combining this database with the 2-anonymous one from HUP, we can again unambiguously infer that Rebecca has HIV.

Individually, the tables satisfy *k*-anonymity, but together, there is a problem. From the first table, we can learn that Rebecca has either HIV or ulcerative colitis. From the second table, we learn that Rebecca has either HIV, lupus, or a hip fracture. But when the tables are taken together, we know with certainty that she has HIV. So there are two major flaws with k-anonymity: it tries to protect against only a very narrow view of what a privacy violation is—an explicit reidentification of patient records—and its guarantees are brittle under multiple data releases.

Another tempting first thought, if our goal is to prevent reidentification, is that the solution to private data analysis is simply not to release any data at the individual level. We should restrict ourselves to releasing only aggregate data—averages over many people, say, or predictive models derived using machine learning. That way, there is nothing to reidentify. But this turns out to be both overly restrictive and again insufficient. Aggregating data turns out to pose plenty of privacy risks, even as it limits the utility of the data. Consider the following example, which shocked the genetics community when it was first discovered in 2008.

The human genome is encoded via a sequence of roughly three billion base pairs—complementary pairs of nucleobases that form the basic building blocks of DNA. Any two people will have genomes that are identical in almost all positions—more than 99 percent of them. But there are certain positions in the genome that may differ between two individuals. The most common forms of genetic variation are called single nucleotide polymorphisms, or SNPs (pronounced "snips"). A SNP represents a position in the genome in which one person might have a certain base pair but another person might have a different one. There are roughly ten million SNPs in the human genome. SNPs can be useful in identifying the genetic causes of disease. The goal of a genome-wide association study (GWAS) is often to find correlations between the presence of genetic variants (alleles) in SNPs and the prevalence of a disease. Generating a basic form of GWAS data might involve gathering a thousand patients who all have some disease—say, pancreatic cancer—sequencing their DNA, and publishing the average allele frequency in each SNP across the entire population. Note that this form of data consists only of averages or statistics: for example, that in a certain SNP position, 65 percent of the population have the C nucleotide and 35 percent have the A nucleotide instead. But because there are so many SNPs, there can be lots of these averages in a single dataset—hundreds of thousands or millions.

The 2008 paper showed that by combining a huge number of simple tests for correlation (one for each SNP), it was possible to test whether a *particular individual's* DNA had been used in computing the average allele frequencies in a particular GWAS dataset—or in other words, whether a particular individual was part of the group of people from whom the GWAS data was gathered. This is not a reidentification, as only one bit of information has been learned: whether the individual was in the group or not. But this is nevertheless a significant privacy concern, because pools of subjects from which GWAS data is gathered often share some disease trait. Learning that an individual was in the GWAS group can reveal that he has, say, pancreatic cancer.

In response to this study, the National Institutes of Health immediately removed all aggregate GWAS data from open-access databases. This mitigated the privacy problem but erected a serious barrier to scientific research of societal importance. More recent research has shown that many other aggregate statistics can leak private data. For example, given only input-output access to a trained machine learning model, it is often possible to identify the data points that were used in its training set. Roughly speaking, this is because machine learning models will at least slightly "overfit" the particular examples they were trained on— for example, a trained model will have higher confidence when asked to classify photos that were used to train it, compared to the confidence it will have when asked to classify examples it has not previously seen.

BREACHES AND INFERENCES

In general, notions of data privacy based on anonymization are susceptible to the serious flaws we have outlined. It might be tempting to instead think that the powerful toolkit of modern cryptography could play an important role in privacy. But it turns out that cryptography solves a different problem, which we might more accurately call data security, rather than providing the type of privacy we're looking for. It's instructive to understand why.

Imagine a simple computational pipeline for medical research. It begins with a database of patient medical records, and our goal is to use machine learning to build a model predicting whether patients have a particular disease based on their observed symptoms, test results and medical history. So the database is given as input to a machine learning algorithm like backpropagation, which in turn outputs the desired predictive neural network.

In this pipeline, we of course want the raw database to be secured. Only authorized personnel—doctors and the researchers building the predictive model—should be allowed to read or alter the database. We want to prevent outright *breaches* of the data (such as the major ones that have occurred at companies like Equifax, Marriott, and Yahoo in recent years). This is the core privacy problem that cryptography attempts to solve, in the form of file encryption algorithms. A useful metaphor is a lock and keys—when the database is locked, only those with the keys should be able to unlock it.

The neural network, however, is different from the original data. We *want* it to be published, released and used—not only by the authorized doctors and researchers, but really by almost anyone. A natural outcome would be for the researchers to publish the details of their model in a scientific journal, so that other researchers and doctors can understand the interactions between various symptoms and the disease in question. Encrypting the neural network, or preventing its use in the field, would obviate the purpose of building it in the first place. In general the purpose of doing computations on data, even including sensitive private data, is to release at least some broad properties about that data to the world at large.

While we are not concerned with unauthorized breaches of the neural network (releasing it was the whole point), we should be concerned about unwanted *inferences* that might be made from it. In particular, we wouldn't want the release of the neural network to allow someone to determine specific details about your medical record, such as whether you have the disease in question. (In fact, recent research has discovered that it often is possible to do exactly that—extract the training data just from access to the learned model.) This is a more nuanced concern than simply locking data down—we want the results of our algorithms to release *useful* information, but not to leak *private* information.

This is the notion of privacy we are concerned with here. And its importance has been rapidly amplified by the era we live in, in which sensitive consumer data is used to build predictive models that are then "released," in the sense that they are used by a very large and diffuse set of entities such as apps, employers, advertisers, lenders, insurers, judges and others. While the design of cryptographic algorithms has been studied for centuries (and experienced colossal scientific and practical advances beginning in the 1970s, with the introduction of so-called public key cryptography), the more subtle inferential privacy we seek is in its relative infancy.

So there are privacy risks beyond simple reidentification. Limiting ourselves to aggregate statistics doesn't fix the problem. And privacy (at least of the type we are seeking) is not the problem that cryptography solves. One way to start thinking rigorously about what to do about all this is to first consider what risks we want to mitigate with data privatization. Perhaps we can start by being ambitious. Can we ask that performing a data analysis should have no risks at all for the people whose data is involved in the analysis? If we could do this, it would seem to be a great privacy definition—the ability to promise complete freedom from harm. We will ultimately see how we can usefully achieve something that is almost as strong—but the following thought experiment illustrates that we will have to refine our goals a bit, and limit our ambitions.

SMOKING MAY BE HARMFUL TO YOUR PRIVACY

Imagine a man named Roger—a physician working in London in 1950 who is also a cigarette smoker. This is before the British Doctors Study, which provided convincing statistical proof linking tobacco smoking to increased risk for lung cancer. In 1951, Richard Doll and Austin Bradford Hill wrote to all registered physicians in the United Kingdom and asked them to participate in a survey about their physical health and smoking habits. Two-thirds of the doctors participated. Although the study would follow them for decades, by 1956 Doll and Hill had already published strong evidence linking smoking to lung cancer. Anyone who was following this work and knows Roger would now increase her estimate of Roger's risk for lung cancer simply because she knows both that he is a smoker and now a relevant fact about the world: that smokers are at increased risk for lung cancer. This inference might lead to real harm for Roger. For example, in the United States, it might cause him to have to pay higher health insurance premiums—a precisely quantifiable cost.

In this scenario, in which Roger comes to harm as the direct result of a data analysis, should we conclude that his privacy was violated by the British Doctors Study? Consider that the above story plays out in exactly the same way even if Roger was among the third of British physicians who declined to participate in the survey and provide their data. The effect of smoking on lung cancer is real and can be discovered with or without any particular individual's private data. In other words, the harm that befell Roger was not because of something that someone identified about his data per se but rather because of a general fact about the world that was revealed by the study. If we were to call this a privacy violation, then it would not be possible to conduct any kind of data analysis while respecting privacy—or, indeed, to conduct science, or even observe the world around us. This is because any fact or correlation we observe to hold in the world at large may change our beliefs about an individual if we can observe one of the variables involved in the correlation.

How, then, should we refine our goals so as to distinguish between protecting the secrets specific to Roger's data while still allowing the discovery of facts about the world that might nevertheless cause people to think differently about Roger? Our first attempt, which un-helpfully would have declared the British Doctors Study a violation of Roger's privacy, can be viewed as comparing the world as it actually is with the imaginary world in which the British Doctors Study was never carried out. It declares Roger's privacy to have been violated because Roger was demonstrably better off in the imaginary world, as compared to the real world. This view of privacy is consistent with what Tore Dalenius defined in 1977 as a goal for statistical database privacy: that nothing about an individual should be learnable from a dataset if it cannot also be learned without access to the dataset. But this thought experiment didn't get to the heart of what we want privacy to mean here—a measure of the harm that might befall Roger *as the result of the use of his data*. To capture this nuance, we can consider a slightly different thought experiment.

Suppose we compare the following two worlds. In the first world, the British Doctors Study is carried out, and Roger opts in to it, so the study includes his data. In the second world, the British Doctors Study is still carried out, but this time Roger opts out of it—so in this world, his data has no effect on the outcome. In both cases, the participation of everyone other than Roger is fixed; the only difference between the two worlds is whether Roger's data was used in the study or not. What if we declare the study to be privacy-preserving if we can promise that Roger is no worse off if the study is performed with his data, compared to if it is performed without his data? In other words, we ask for a refinement of Dalenius's goal: that nothing about an individual should be learnable from a dataset that cannot be learned from the *same dataset but with that individual's data removed*. This kind of definition still has the promise that Roger cannot be harmed by the use of his data, and it might plausibly allow useful science such as the British Doctors Survey to be carried out. In that scenario, Roger's insurance rates were raised because of the discovered link between smoking and lung cancer—but that fact about the world would have been discovered whether or not he opted into the study.

A DIFFERENT(IAL) NOTION OF PRIVACY

Enter differential privacy, a stringent measure of privacy that still allows us to gain useful insights from data. Differential privacy is a mathematical formalization of the foregoing idea—that we should be comparing what someone might learn from an analysis if any particular person's data was included in the dataset with what someone might learn if it was not. The concept was developed in the early 2000s by a team of theoretical computer scientists who were subsequently awarded the prestigious Gödel Prize: Cynthia Dwork, Frank McSherry, Kobbi Nissim, and Adam Smith. To describe what differential privacy asks for, it is important to first understand what a randomized algorithm is.

Let's start with an algorithm. Remember, this is just a precise description of how to take inputs to some problem and process them to arrive at the desired outputs. In the introduction, we saw an example of a sorting algorithm. But an algorithm might also take as input health records and map them to a set of features that appear to be correlated. Or it might take as input video rental records and map them to a set of movie recommendations for each customer. The important thing is that the algorithm precisely specifies the process by which the inputs are mapped to the outputs. A randomized algorithm is just an algorithm that is allowed to use randomization. Think about it as an algorithm that can flip coins as part of its precisely specified process and then make decisions that depend on the outcome of the coin flips. So a randomized algorithm maps inputs to the probabilities of different outputs. (We'll see a concrete example of a simple randomized algorithm in the next section.)

The introduction of randomness into algorithms has varied and powerful uses, including for the generation of cryptographic keys, for speeding up the search for solutions of algebraic equations, and for balancing out the loads on a distributed collection of servers. The use of randomness in differential privacy is for yet another purpose–namely, to deliberately add *noise* to computations, in a way that promises that any one person's data cannot be reverse-engineered from the results.

Differential privacy requires that adding or removing the data record of a single individual not change the probability of any outcome by "much" (we'll explain what this means in a minute). It requires this of an algorithm even in the worst case, no matter what records the other individuals have provided and no matter how unusual the added or removed data is. And differential privacy is a constraint that comes with a tunable "knob" or parameter, which can be thought of as a measure of the amount of privacy demanded. This knob governs how much a single individual's data can change the probability of any outcome. For example, if the privacy knob is set to 2.0, then differential privacy requires that no outcome be more than twice as likely if the algorithm is run with Roger's data compared with the case in which it is run on the same dataset but with Roger's data removed.[1]

Let's consider for a moment why the mathematical constraint of differential privacy corresponds to something we might think of as a

---

[1] We are actually describing the exponential of the privacy parameter, as usually defined in the mathematical literature. If we were to be consistent with the mathematical literature, we would say that the privacy parameter was $\ln(2)$ if changing Roger's data could cause an event to double in probability.

notion of privacy. We'll give three interpretations here, but there are more.

Most basically, differential privacy promises safety against arbitrary harms. It guarantees that no matter what your data is, and no matter what thing you are concerned about occurring because of the use of your data, that thing becomes (almost) no more likely if you allow your data to be included in the study, compared to if you do not. It literally promises this about anything you can think of. It promises that the probability that you get annoying telemarketing calls during dinner does not increase by very much if you allow your data to be included in a study. It promises that the probability that your health insurance rates go up does not increase by very much if you allow your data to be included in a study. And it certainly promises that the probability that your data record is reidentified (as in the Massachusetts hospital record and Netflix Prize examples) does not increase by very much. The strength of the promise of differential privacy is that it does not matter what harm you are worried about—differential privacy promises that the risk of *any* harm does not increase by more than a little bit as the result of the use of any individual's data. And it makes sense even in settings in which there is no data to reidentify, as in the GWAS (genomics) example.

But what if privacy is about what others can learn about you, instead of what bad things can happen to you? Differential privacy can also be interpreted as a promise that no outside observer can learn very much about any individual because of that person's specific data, while still allowing observers to change their beliefs about particular individuals as a result of learning general facts about the world, such as that smoking and lung cancer are correlated.

To clarify this, we need to think for a moment about how learning (machine or otherwise) works. The framework of Bayesian statistics provides a mathematical formalization of learning. A learner starts out with some set of initial beliefs about the world. Whenever he observes something, he changes his beliefs about the world. After he updates his beliefs, he now has a new set of beliefs about the world (his posterior beliefs). Differential privacy provides the following guarantee: for every individual in the dataset, and for any observer no matter what their initial beliefs about the world were, after observing the output of a differentially private computation, their posterior belief about anything is close to what it would have been had they observed the output of the same computation run without the individual's data. Again, "close" here is governed by the privacy parameter or knob.

As one final way of interpreting this same privacy guarantee, suppose some outside observer is trying to guess whether a particular person—say, Rebecca—is in the dataset of interest or not (or whether her record specifies some particular ailment, such as lung cancer, or not). The observer is allowed to use an arbitrary rule to make his guess, based on the output of the differentially private computation. If the observer is shown either the output of a computation run with Rebecca's data or the output of the same computation run without her data, he will not be able to guess which output he was shown substantially more accurately than random guessing.

The three interpretations provided above are really just three different ways of looking at the same guarantee. But we hope they will convince you of what they have convinced many scientists: that differential privacy is among the strongest kinds of individual privacy assurances we could hope to provide without a wholesale ban on any practical use of data (for example, never doing valuable studies like the British Doctors Study linking smoking and lung cancer). The main question is whether it might be *too* strong, presenting such an onerous constraint that, for example, it is incompatible with modern machine learning. As we shall see, there is a rather happy middle ground between differential privacy and machine learning (and many other types of computations).

HOW TO CONDUCT EMBARRASSING POLLS

Differential privacy promises individuals a very strong kind of protection. But it would not be interesting if it were impossible to achieve while doing any kind of useful data analysis. Fortunately, this is not the case. In fact, it is possible to conduct essentially any kind of statistical analysis subject to differential privacy. But privacy doesn't come for free: one will generally need more data to obtain the same level of accuracy than one would need without a privacy constraint, and the more stringently one sets the privacy parameter, the more serious this tradeoff becomes.

To see an example of why this is possible, let's consider the simplest of all possible statistical analyses: computing an average. Suppose we want to conduct a poll to find out how many men in Philadelphia have ever cheated on their wives. A straightforward way to do this would be to attempt to randomly sample a reasonable number of men from the population of Philadelphians, call them, and ask them if they have ever had an affair. We would write down the answer provided by each one. Once we collected all of the data, we would enter it into a spreadsheet and compute the average of the responses, and perhaps some accompanying statistics (such as confidence intervals or error bars). Note that although all we wanted was a statistic about the population, we would have incidentally collected lots of compromising information about specific individuals. Our data could be stolen or subpoenaed for use in divorce proceedings, and as a result, people might rightly be concerned about participating in our poll. Or they might simply be embarrassed to admit to having had an affair, even to a stranger.

But consider the following alternative way of conducting a poll. Again, we would randomly sample a reasonable number of men from the population of Philadelphians (somewhat more than we needed before). We would call them up and ask them if they have ever had an affair. But rather than asking them to answer our question directly, we would give them the following instructions: Flip a coin, and don't tell us how it landed. If it came up heads, tell us (honestly) whether you have ever cheated on your wife. But if it came up tails, tell us a *random* answer: flip the coin again and tell us yes if it comes up heads and no if it comes up tails. This polling protocol is an example of a simple randomized algorithm.

The result is that when we ask people whether they have ever had an affair, three-quarters of the time they tell us the truth (half the time the protocol tells them to tell us the truth, and if instead the protocol tells them to respond with a random answer, then half of *that* time they just happen to tell us the right answer at random). The remaining one-quarter of the time they tell us a lie. We have no way of telling true answers from lies. We record the answers that they give us—but now everyone has a strong form of plausible deniability. If in a divorce proceeding our records are subpoenaed and it is revealed that a man in our survey answered yes to our survey protocol, the accused husband can reasonably protest that in fact he had never had an affair—the coins simply came up such that he was asked to randomly report yes. Indeed, nobody can form strong beliefs about the true data of any single individual when records are collected in this way.

Yet from data collected in this way, it is still possible to get a highly accurate estimate of the fraction of Philadelphia men who have cheated on their wives. The key is that although the individual answers we are collecting are now highly error-prone, we know exactly how those errors are introduced, and so we can work backward to remove them, at least in the aggregate. Suppose, for example, that it happens to be that one-third of men in Philadelphia have at some point cheated on their wives. Then how many people in our survey do we expect to answer yes? We can compute this, since we know that people answer truthfully three-quarters of the time. One-third of the people in the

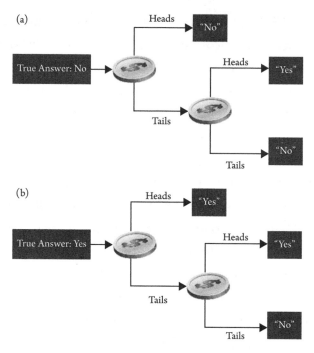

Fig. 3. If your true answer is no, the randomized response answers no two out of three times. It answers no only one out of three times if your answer is yes.

survey have a truthful answer of yes, and three-quarters of those will report yes, for a total of $1/3 \times 3/4 = 1/4$ of the population. Additionally, two-thirds of the people in the survey have a truthful answer of no, and one-quarter of these will report yes, for a total of $2/3 \times 1/4 = 1/6$. In total, we expect $1/4 + 1/6 = 5/12$ of the population to report yes in this case.

Because we know this, if we observe that five-twelfths of our survey population have reported yes, we can work backward to deduce that approximately one-third of male Philadelphians have cheated on their wives. This holds for any other proportion we observe—since we know the process by which errors have been introduced, we can work backward to deduce approximately the fraction of the population for whom the truthful answer is yes. This process is approximate, because even if exactly one-third of the men we survey have cheated on their wives, this only tells us that five-twelfths of the population will report yes *on average*. The actual fraction will deviate slightly from five-twelfths because of how individuals' coin flips actually turn out. But because we are now reasoning about an average computed over a large sample, it will be very accurate—and will get more accurate the more people we survey. This is exactly the same effect we see when we flip coins. If we flip ten coins, we might expect the fraction of heads we see in total to deviate substantially from one-half. But if we flip ten thousand coins, we expect the fraction of heads that we see to be very close to one-half indeed. In exactly the same way, the error introduced in our survey by the randomness that was added for privacy shrinks to zero as we include more and more people in our survey. This is just an instance of what is known as the "law of large numbers" in statistics.

Although this protocol is simple, the result is remarkable: we are able to learn what we wanted without incidentally collecting any strongly incriminating information about any single individual in the population. We managed to skip this middle step and go straight to collecting only the aggregate information that we wanted.

The randomized polling protocol we have just described is old—it is known as randomized response, and it dates to 1965, decades before the introduction of differential privacy. But it turns out that the protocol satisfies 3-differential privacy—that is, differential privacy in which any bad event you were worried about is at most three times more likely if you participated in the survey.[2] There is nothing special about the number 3 here, though. By decreasing the probability that the protocol asks people to tell the truth, we can make the protocol more private. By increasing the probability, we can make the protocol less private. This lets us quantitatively manage an important trade-off:

---

[2] In the standard accounting, we would say it satisfies $\ln(3)$ differential privacy, where ln denotes natural logarithm.

the more private we make the protocol, the more plausible denia-bility each individual we poll gets, and the less anyone can learn from an individual polling response. On the other hand, the smaller the privacy parameter gets, the more error-prone the process of working backward gets—mapping from the proportion of yes an-swers in our poll to the actual proportion of cheaters in our popula-tion. So to get the same accuracy, the smaller the privacy parameter, the more people we need to poll.

Computing an average is the simplest statistical analysis you might want to do—but it is by no means the only one you can do with dif-ferential privacy. The biggest strength of differential privacy is that it is compositional, which means it doesn't break the way k-anonymity does when you run more than one differentially private analysis. If you've got two differentially private algorithms, you can run them both and the result is still differentially private. You can give the output of one as the input to the other, and the result is still differentially pri-vate. This is enormously useful, because it means you can build com-plicated algorithms by piecing together simple differentially private building blocks. It would be a tedious mess if we needed to reason about the privacy properties of every new algorithm afresh—and it would be a horrendously complex problem if the algorithm was long and complicated.

Instead, we can reason about the privacy guarantees of simple com-ponents, such as computing an average. We can then go about de-signing complicated algorithms by gluing together these simple primitives in various ways, and be guaranteed to still satisfy differen-tial privacy. This makes private algorithm design modular, just like standard nonprivate algorithm design. So we can go from computing averages to performing some optimization over a dataset and then to privately training neural networks. As a general rule of thumb, any statistical analysis or optimization—one whose success is defined with respect to the underlying distribution, and not with respect to any particular dataset—can be made differentially private, albeit usually with a need for more data.

WHOM DO YOU TRUST?

Randomized response actually satisfies an even stronger constraint than differential privacy. Differential privacy only requires that nobody should be able to guess substantially better than randomly whether a given output was computed with an individual's data or without it. In our polling example, the natural output of the computation is the final average we compute: the estimated proportion of Philadelphia husbands who have cheated on their wives. But randomized response promises not only that the final average is differentially private but also that the entire set of collected records is as well. In other words, randomized response satisfies differential privacy even if we publish the pollster's entire spreadsheet, with all the data she has collected, and not just the final average. It promises differential privacy not just from an outside observer but from the pollster herself.

If we only wanted differential privacy from an outside observer (and trusted the pollster not to leak the raw data), then she could have conducted the poll in the standard way and only added noise just to the final average that she published. The benefit of this approach is that far less noise would be necessary. In randomized response, every person individually adds enough noise to obscure his or her data point. When we average all of the provided data, in aggregate we have added much more noise than we really needed. If we trust the pollster with the real data, she can aggregate the data and then add just enough noise to obscure the contribution of one person. This leads to a more accurate estimate, but of course it doesn't come for free: it doesn't provide the stronger guarantee of safety if the pollster's records are subpoenaed.

Whether we want this stronger guarantee or not (and whether it is worth the trade-off with error) depends on what our model of the world is and who we think is trying to violate our privacy. The standard guarantee of differential privacy assumes that the algorithm analyzing the data is run by a trusted administrator: that is, we trust the algorithm (and whoever has access to its implementation) to do with our data only what it is supposed to do. The algorithm and the administrator are the ones who add the privacy protections—we trust them to do this. Because we have a trusted administrator who can aggregate all the data "in the clear" (before privacy is added), this is sometimes referred to as the centralized model of differential privacy.

Randomized response instead operates in what is called the local or decentralized model of differential privacy. Here there is no trusted data administrator. Individuals add their own privacy protection locally, in the form of data perturbations, before they hand it over. This is what happens in the randomized response polling method: each person flips his or her own coin and gives a random response to the pollster, and the pollster never sees the true answers. Another suggestive way of thinking about centralized versus local differential privacy is whether the privacy is added on the "server" side (centralized) or the "client" side (local).

In many ways, deciding which trust model you want is a more important decision than setting the quantitative privacy parameter of your algorithm. But the two decisions cannot be made in isolation. Since differential privacy in the local model gives a much stronger guarantee, it is not surprising that choosing it has a substantial cost. In general, for a fixed privacy parameter (say, 2), satisfying differential privacy in the local model will lead to a less accurate analysis than in the centralized trust model. Or, turning this trade-off on its head, for a fixed accuracy requirement, opting for the local model will require either more data than would be needed in the centralized model or else a worse privacy parameter. These realities play a big role in shaping the three most important deployments of differential privacy to date.

OUT OF THE LAB AND INTO THE WILD

Two of the first large-scale commercial deployments of differential privacy were implemented by Google and Apple. In 2014, Google announced on its security blog that its Chrome browser had started collecting certain usage statistics about malware on users' computers with differential privacy. In 2016, Apple announced that iPhones would begin collecting usage statistics using differential privacy. These two large-scale deployments have many things in common. First, both Google and Apple decided to operate in the local trust model, basing their algorithms on randomized-response-like methods. This means that Google and Apple are never collecting the relevant private data itself. Instead, your iPhone is simulating coin flips to run a randomized response protocol and sending to Apple only the random outputs generated. Second, both deployments were used to privately gather data that the companies had previously not been gathering at all, rather than implementing differential privacy on top of datasets that they already had available to them.

For Google and Apple, the trade-offs that go with the local model of privacy make sense. First, neither company is necessarily trusted by its users. No matter what you think about the companies themselves, there is a real risk that data they store on their servers will be revealed to others via hacking or government subpoena. For example, in 2013, Edward Snowden released thousands of documents revealing intelligence-gathering techniques used by the National Security Agency (NSA). Among the revelations were that the NSA had been eavesdropping on communications that flowed between Google (and Yahoo) data centers, without Google's knowledge. Security has been tightened since then—but whether or not surreptitious data exfiltration has been stopped, there is still much data released to national governments via the ordinary legal process. Google reports that in the one-year period starting July 2016, government authorities requested data for more than 157,000 user accounts. Google produced data in response to roughly 65 percent of these requests. A guarantee of differential privacy in the centralized model promises nothing against these kinds of threats: so long as Google or Apple holds customer data, it is subject to release through nonprivate channels.

On the other hand, in the local model of differential privacy, companies such as Google and Apple never have to collect the private data in the first place. Instead, they play the role of the pollster in our example above, only recording the noisy responses of users who always maintain a strong guarantee of plausible deniability. The pollster might lose her spreadsheet to a hacker who breaks into her computer—but the spreadsheet doesn't contain much information about any particular person. A divorce lawyer might subpoena the record of how his client's husband answered the pollster's question—but the answer he gets back doesn't tell him what he wanted to know. Of course, we've seen that asking for this stronger subpoena-proof form of privacy protection comes at a cost: to get an accurate picture of population-wide statistics, you need a *lot* of data. But both Google and Apple are in a strong position to make this trade-off. Google has more than a billion active Chrome users, and Apple has sold more than a billion iPhones.

It is also noteworthy that both Google and Apple applied differential privacy to data that they weren't collecting before, rather than to data sources they already had available. Adding privacy protections can be a tough sell to engineers who already have the data available to them. Asking for privacy almost always means a real, measurable degradation in data quality in exchange for a decrease in risk that can feel much more nebulous. It is hard to convince an engineering team to give up access to a clean data source they already have. But if differential privacy is framed as a way to get access to new data sources— data that previously was not being collected at all, because of privacy concerns—that is a different story entirely. When framed this way, differential privacy is a way to get more data, not an obligation that degrades existing analyses. This is what happened at both Google and Apple.

A third large-scale deployment of differential privacy serves as an interesting contrast to the Google and Apple use cases. In September 2017, the US Census Bureau announced that all statistical analyses published as part of the 2020 Census will be protected with differential privacy. In contrast to the large industrial deployments, the Census Bureau will operate using the centralized model of differential privacy, collecting the data (as it always has) exactly and adding privacy protections to the aggregate statistics it publicly releases. Moreover, this isn't letting the Census Bureau access a new data source—the requirement to conduct a census every ten years is laid out in the US Constitution, and the first census was conducted in 1790.

So why did the Census Bureau decide to adopt differential privacy in the first place? And why did it opt for the higher accuracy that comes with the weaker centralized trust model, which does not protect against subpoenas, hacks, and the like?

The answer to the first question is that the Census Bureau is legally mandated to take measures to protect individual privacy. For example, all Census Bureau employees take an oath of nondisclosure and are sworn for life to protect all information that could identify individuals. There is no option to release information in the clear: privacy is a must, and the only question is what privacy protections should be used. So the alternative wasn't to do nothing about privacy—it was to continue taking the heuristic approaches of prior censuses that had no hard promises about privacy and difficult-to-quantify effects on data accuracy. There is no technology other than differential privacy that gives similarly principled, formal privacy guarantees while preserving the ability to estimate population-wide statistics.

The answer to the second question—about adopting the centralized model—is that legal protections give the Census Bureau a much stronger claim to being a trusted administrator than Google or Apple. By law, the Census Bureau cannot share individual data records—that is, your responses to questions on the 2020 Census—even with any other government agency, including the IRS, the FBI, or the CIA.

WHAT DIFFERENTIAL PRIVACY DOES NOT PROMISE

At its core, differential privacy is meant to protect the secrets held in individual data records while allowing the computation of aggregate statistics. But some secrets are embedded in the records of *many* people. Differential privacy does not protect these.

The fitness tracking website Strava is a dramatic example of this. It allows users to upload data from devices such as Fitbits to their website, so that they can keep track of their activity history and locations. In November 2017, Strava released a data visualization of the aggregate activity of its users, consisting of more than three trillion uploaded GPS coordinates. This lets you see, for example, popular running routes in almost every major city. But large parts of the globe are almost devoid of activity on this map—for example, poor and war-torn regions including Syria, Somalia, and Afghanistan. Most people living in these places don't have Fitbits or use Strava.

But there is a notable exception in each of these countries: US military personnel. The US military encourages its soldiers to use Fitbits, and they do. The Strava data reveals (in aggregate) the most popular jogging routes in each of these locations, as it is intended to do. But the most popular jogging routes in Afghanistan's Helmand Province turn out to be on military bases, not all of which are supposed to be publicly known. The Strava data revealed sensitive state secrets, even though it did not necessarily reveal any individual secret. Data such as the Strava location heatmap could have been assembled with differential privacy protections, but this would only promise that the map would be nearly unchanged if any single soldier had decided not to use his Fitbit—it does *not* promise that the aggregate behavior of soldiers on a military base would be hidden. (It's worth noting that differential privacy does allow the data of small groups to be protected—in particular, an algorithm promising 3-differential privacy for individuals also provides 3^k-differential privacy for the data of sets of k individuals—but if k is large, as with an entire platoon of soldiers, this won't be a very meaningful guarantee.)

Differential privacy can also allow people to learn about secrets that you think of as your own, if it turns out that they can be deduced from publicly available data—often in ways that you could not anticipate. Remember that by design, differential privacy protects you from the consequences of what an observer might be able to learn specifically because of your data but does not protect you from what an attacker might be able to infer about you using general knowledge of the world. So it prevented an observer from deducing that Roger had lung cancer as a result of his participation in the British Doctors Survey, but it didn't prevent us from learning that smoking was a good predictor of lung cancer—even though this might cause us to change our beliefs about Roger's risk for lung cancer if we know he is a smoker. In general this is a good thing—it lets us still learn about the world and conduct science while being able to meaningfully protect privacy.

But the world is full of curious correlations, and as machine learning gets more powerful and data sources become more diverse, we are able to learn more and more facts about the world that let us infer information about individuals that they might have wanted to keep private. In the simple example of the British Doctors Survey, if only Roger knew the conclusion of the study ahead of time, he could have tried to keep his smoking habits private. But how was he to know? Similarly, we make all sorts of seemingly innocuous information about ourselves public, but this opens us up to a surprisingly large number of inferences.

For example, Facebook offers many options allowing us to either hide or reveal information about ourselves on our Facebook profiles.

If we want, we can decide not to show our sexual orientation, our relationship status, or our age on our public profiles. This gives users a sense of control. But by default, Facebook makes public what pages we "like." For example, you might like pages for *The Colbert Report*, curly fries, and Hello Kitty. This seemingly trivial stream of data gives your user profile on Facebook a statistical fingerprint that can be correlated with other things. In 2013, a team of researchers at Cambridge University showed that this stream of data could be mined to uncover correlations that would accurately map people's seemingly innocuous likes to properties about them that they may have wanted to keep private. From a user's likes, the researchers were able to guess at the user's gender; political affiliation (Republican or Democratic); sexual orientation; religion; relationship status; use of drugs, cigarettes, and alcohol; and even whether the user's parents divorced before he or she turned twenty-one—all with a statistically significant level of accuracy. These are all things that you might have wanted to keep private, or at least not post publicly on the Internet for everyone to see. But you probably didn't worry about hiding that you like curly fries.

In a similar demonstration in 2010, a machine learning start-up called Hunch released a demo that it called the Twitter Predictor Game. It was simple: you told it your Twitter handle, and it scoured the Twitter network to gather some simple data, consisting of whom you followed and who followed you. From this apparently innocuous data, it then asked you a series of private questions—things you likely never tweeted about (it didn't read your tweets anyway)—and it would guess the answers. It could accurately guess whether you were pro-choice or pro-life. Whether you thought gay marriage should be legal. Whether you had a brokerage account. Whether you listened to talk radio. How often you wore fragrance. Whether you had ever purchased something from an infomercial. Whether you preferred Star Wars or Star Trek. Whether you had ever kept a journal.

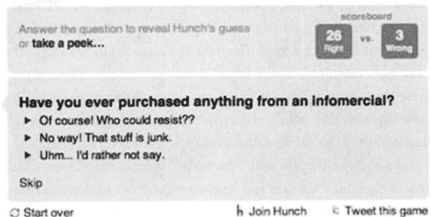

hunch thinks you'll say:

Answer the question to reveal Hunch's guess or **take a peek...**

scoreboard

26 Right vs. 3 Wrong

Have you ever purchased anything from an infomercial?
▶ Of course! Who could resist??
▶ No way! That stuff is junk.
▶ Uhm... I'd rather not say.

Skip

↻ Start over   ♮ Join Hunch   ╚ Tweet this game

Fig. 4. Hunch would ask users a wide variety of questions and then predict their answers from their pattern of followers on Twitter.

It was shockingly accurate—it correctly predicted user answers about 85 percent of the time. It wasn't exploiting your private data to answer these questions, because you never gave it access to data explicitly answering those questions. Data explicitly answering those questions didn't exist. Instead, it was looking only at data you had made public—the list of people you followed on Twitter—and making inferences from information it had about the population at large. It was doing social science on a large scale—it was learning about the world—and by design, differential privacy is not something that would protect you from people making these kinds of inferences about you.

These two examples are nothing more than the modern version of knowing that smoking increases one's risk of lung cancer and then making inferences about the cancer risk of a known smoker. But as machine learning becomes more powerful, we are able to discover increasingly complicated and non-obvious correlations. Seemingly useless bits of data—the digital exhaust that everyone is constantly producing as they use the Internet, purchasing items, clicking on links, and liking posts—turn out to be surprisingly accurate predictors of other things that we might really wish that other people didn't know. Differential privacy doesn't protect you from these kinds of inferences, because the correlations they rely on would exist whether or not your data was available. They are just facts about the world that large-scale machine learning lets us discover. And this is by design— these kinds of privacy risks will always exist whenever people are allowed to study the world and draw conclusions. There isn't any way to stop it, short of hiding all data and stopping science altogether.

Let's end the chapter with one more example that demonstrates how facts about the world and facts about you can sometimes be entangled. Your DNA represents some of the most sensitive information that there is about you. In a very real sense it encodes your identity: what you look like, where your ancestors came from, and what diseases you might have or be susceptible to. And it can also be used as forensic evidence to place you at the scene of a crime, and ultimately to convict and imprison you. But your DNA is not yours alone: you share much of it with your parents, your children, your siblings, and, to a lesser extent, all of your relatives up and down your family tree. So although we think of your DNA as *your* data—something you are free to make public or keep secret at your discretion—what you do with it affects other people as well. For example, if you have committed a crime and are on the run from police, you would be foolish to upload your DNA to a publicly searchable database, because the police could match the public record to a sample from the crime scene and apprehend you. But you might not be able to stop your relatives from uploading *their* genetic information.

This is exactly how the infamous Golden State Killer was captured in 2018. He had been linked to more than fifty rapes and twelve murders between 1976 and 1986, leaving plenty of DNA evidence behind— but since he wasn't in any law enforcement database, the trail had gone cold and police had been unable to find him. But times have changed, and law enforcement DNA databases are no longer the only game in town. Starting in the mid-2000s, people began to voluntarily upload their own DNA to public databases on the Internet, to learn more about their family histories. For example, in 2011, two volunteers started a website called GEDmatch, which hobbyists could use to upload DNA profiles that they had generated using commercial sites such as 23andMe. Users could search for partial matches, ostensibly to find their own distant relatives and link their family trees, which GEDmatch made available. But anyone else could also conduct such a search— and police still investigating the Golden State Killer uploaded a sample of his DNA, taken from a crime scene, in hopes of finding a match. And they did—not to the Golden State Killer himself, who hadn't uploaded his own DNA, but to a number of his relatives, who had. From their family trees, the police were able to find a small list of suspects, ultimately arresting Joseph James DeAngelo at age seventy-two.

[GED match].® Tools for DNA & Genealogy Research

April 27, 2018 We understand that the GEDmatch database was used to help identify the Golden State Killer. Although we were not approached by law enforcement or anyone else about this case or about the DNA, it has always been GEDmatch�s policy to inform users that the database could be used for other uses, as set forth in the Site Policy (linked to the login page and https://www.gedmatch.com/policy.php). While the database was created for genealogical research, it is important that GEDmatch participants understand the possible uses of their DNA, including identification of relatives that have committed crimes or were victims of crimes. If you are concerned about non-genealogical uses of your DNA, you should not upload your DNA to the database and/or you should remove DNA that has already been uploaded.To delete your registration contact gedmatch@gmail.com

Fig. 5. A message posted on the GEDmatch website to its users after the Golden State Killer was arrested thanks to DNA data found on its website.

The case of the Golden State Killer raises a number of difficult privacy challenges, because genetic data brings into sharp relief two related things we think of as personal rights: the autonomy to do what we like with "our own" data, and the ability to control our "private" data. Because your DNA contains information not just about you but about your relatives, these two things are in conflict: your "private" data is not always "your own." As Erin Murphy, a law professor at New York University, told the *New York Times*, "If your sibling or parent or child engaged in this activity online, they are compromising your family for generations." And because of this, this is not the kind of privacy threat that differential privacy is designed to protect against. Differential privacy controls the difference between what can be learned about you when you make the individual decision to allow your data to be used and what can be learned when you don't. So it gives you little incentive to withhold your data. But this might be because, to take the case of a genetic database, much about you would become known even if you did withhold your data, because some of your relatives freely provided theirs. This kind of reasoning about how the effect of your own actions in isolation affects incentives is the domain of game theory, and the topic of Chapter 3. Because differential privacy is based on the same basic idea, it might not come as a surprise that tools from differential privacy will be useful in game theory as well.

2

Algorithmic Fairness

From Parity to Pareto

BIAS BY ANALOGY

If you're sufficiently beyond your high school days, you might remember—with love or loathing—the multiple-choice word analogy problems that were formerly a staple of standardized testing. Such problems ask one to consider relationships such as "*runner* is to *marathon*..." and realize that the correct response is "... as *oarsman* is to *regatta*" and not, for instance, "... as *martyr* is to *massacre*." These linguistic puzzles were removed from the SAT in 2005, in part over concerns that they were biased in favor of certain socioeconomic groups—for example, those who knew what regattas were. Perhaps because the loathers outnumbered the lovers, word analogies and their cultural biases seemed to have been forgotten for over a decade.

But they resurfaced in 2016 in a new context, when a team of computer science researchers subjected Google's publicly available "word embedding" model to a clever test based on word analogies. The idea behind a word embedding is to take a colossal collection of text and compute statistics about the so-called co-occurrences of words. For instance, the words *quarterback* and *football* are more likely to be found in close proximity to each other in a document, paragraph, or sentence than the words *quarterback* and *quantum*. These pairwise co-occurrence statistics are then fed to an algorithm that attempts to position (or embed) each word in 2-, 3- or higher-dimensional space in such a way that the distances between pairs of words approximately reflect their co-occurrence statistics. In this sense, words that are more "similar" (as reflected entirely by their empirical usage in the documents) are placed "nearer" each other.

Building a large-scale word embedding is an exercise in data collection, statistics, and algorithm design—that is, it is a machine learning project. Google's "word2vec" (shorthand for "word to vector") embedding is one of the best-known open-source models of its kind and has myriad uses in the many language-centric services Google provides—for instance, realizing that *bike* may be a synonym for *bicycle* when in proximity to *mountain*, but for *motorcycle* when in proximity to *Harley-Davidson*.

The starting point behind the 2016 work was an observation dating back to the 1970s: if we have a good embedding of words into, say, 2-dimensional space, then word analogies should roughly correspond to parallelograms. Thus if *man* really is to *king* as *woman* is to *queen*, then the four points corresponding to these words in the embedding should define two sets of parallel lines—namely, the *man-woman* and *king-queen* pair, and the *king-man* and *queen-woman* pair.

We can use this observation to "solve" for the missing word in analogy problems. For if we ask questions of the form "*Runner* is to *marathon* as *oarsman* is to what?," the three specified words—in this case, *runner*, *marathon*, and *oarsman*—define three corners of a

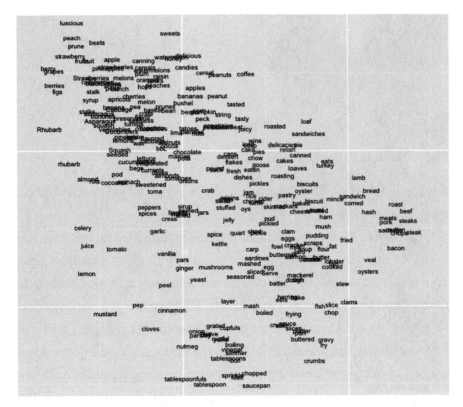

Fig. 6. A small word embedding in two dimensions.

parallelogram in the word embedding, which in turn determines where in space the fourth corner should lie. By looking at the word lying closest to this missing corner, we find the "what"—the word that the embedding "thinks" is the best completion of the analogy. (See Figure 7.)

The team of researchers could have stopped there and used this observation to see how well word2vec would have fared on 1990s SAT tests. But just as with people, sometimes the failings of algorithms and models are far more revealing and interesting than their strengths. So the researchers deliberately restricted the word analogies they investigated to those of the form "*Man* is to *X* as *woman* is to *Y*," where *X* was specified (giving the required three corners) but *Y* (the missing corner) was not—as in "*Man* is to *computer programmer* as

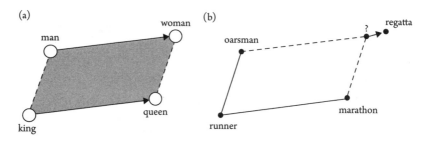

Fig. 7. Word analogies as parallelograms.

woman is to what?" In other words, they specifically tested word2vec on gender analogies.

The findings were dramatic, and demonstrated that word2vec was guilty of rampant gender bias and stereotyping. The example given above was answered by the title of the research article: "Man Is to Computer Programmer as Woman Is to Homemaker?" The paper documented the systematic manner in which word2vec reflected, and perhaps amplified, the biases already present in the raw documents it was trained upon. In the small sample of the embedding reproduced below in Figure 8, one can even visually complete analogies of this form. For instance, it would seem that *ladies* is to *earrings* as *nephew* is to *genius*. While this makes little sense linguistically, it nevertheless feels sexist, with women being associated with ornamentation and men with brilliance. While word analogies may seem a bit esoteric, the important point is that word embeddings are used as basic building blocks for more complicated learning algorithms. And more serious problems can arise when word embeddings and other biased models are used as components in more consequential applications.

In fact, in late 2018, something along these lines was discovered in the machine learning model that Amazon was building to evaluate the resumes of candidates for software engineering jobs. Its algorithm was found to be explicitly penalizing resumes that contained the word *women's*, as in "women's chess club captain," and downgraded candidates who listed the names of two particular all-women colleges. Amazon ultimately disbanded the team working on this particular project, but not before some damage was done—at least in the form of bad publicity for Amazon, and potentially in the form of implicitly enabling discriminatory hiring.

Neither the authors of the word embedding article nor any of its discussants suggested that the bias of word2vec was the result of sexist programmers at Google, unrepresentative or corrupted data, or coding errors. And there is also no reason to suspect that bias in Amazon's hiring tool was the result of malice either. Any of those explanations might have been more reassuring than the truth, which is that the bias was the natural if unexpected outcome of professional scientists and engineers carefully applying rigorous and principled machine learning methodology to massive and complex datasets.

The problem here is that the training data used in machine learning applications can often contain all kinds of hidden (and not-so-hidden) biases, and the act of building complex models from such data can both amplify these biases and introduce new ones. As we discussed in the introduction, machine learning won't give you things like gender neutrality "for free" that you didn't explicitly ask for. Thus even though probably very few of the documents used to create the word embedding,

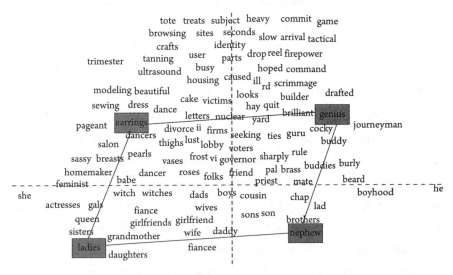

Fig. 8. A word embedding exhibiting gender bias.

if any, exhibited blatant sexism (and certainly none of them actually suggested that homemaker was the best female analogue for male computer programmer), the tiny collective forces of language usage throughout the entire dataset, when compressed into a predictive model for word analogies, resulted in clear gender bias. And when such models then become the basis for widely deployed services such as search engines, targeted advertising, and hiring tools, the bias can be further propagated and even amplified by their reach and scale. This is an example of the complex feedback loops fueled by machine learning that we will examine in Chapter 4.

The embedding bias paper deservedly received a great deal of both academic and media attention. But this is only one example of the growing number of cases in which algorithms—and, most notably, algorithms based on models derived from data via machine learning—exhibit demonstrable bias and discrimination based on gender, race, age, and perhaps many other as yet unknown factors and combinations of factors. And while bias in the search results or ads we are shown may seem (perhaps incorrectly) like a relatively low-stakes affair, the same problems arise in much more obviously consequential domains, such as criminal sentencing, consumer lending, college admissions, and hiring.

As we said in the introduction, many of the problems we identify in this chapter have been discussed well at a high level in other places. Our particular interest will focus on the aspects of these problems that are specific to machine learning, and especially to potential solutions that are themselves algorithmic and on firm scientific footing. Indeed, while the main title of the word embedding paper expressed alarm over the problem uncovered, it also had a more optimistic subtitle: "Debiasing Word Embeddings." The paper reveals a serious concern, but it also suggests a principled algorithm for building models that can avoid or reduce those concerns. This algorithm again uses machine learning, but this time to distinguish between words and phrases that are inherently gendered (such as *king* and *queen*) and those that are not (such as *computer programmer*). By making this distinction, the algorithm is able to "subtract off" the bias in the data associated with nongendered words, thus reducing analogy completions like the one in the paper's title, while still preserving "correct" analogies like "*Man* is to *king* as *woman* is to *queen*."

These are the themes of this chapter: scientific notions of algorithmic (and human) bias and discrimination, how to detect and measure them, how to design fairer algorithmic solution—and what the costs of fairness might be to predictive accuracy and other important objectives, just as we examined the costs to accuracy of differential privacy. We will eventually show how such costs can be made quantitative in the form of what are known as *Pareto curves* specifying the theoretical and empirical trade-offs between fairness and accuracy.

But ultimately, science can only take us so far, and human judgments and norms will always play the essential role of choosing where on such curves we want society to be, and what notion of fairness we want to enforce in the first place. Good algorithm design can specify a menu of solutions, but people still have to pick one of them.

LEARNING ALL ABOUT YOU

In machine learning, word embeddings are an example of what is known as unsupervised learning, where our aim is not to make decisions or predictions about some particular, prespecified outcome but simply to find and visualize structure in large datasets (in this case, structure about word similarity in documents). A far more common type of machine learning is the supervised variety, where we wish to use data to make specific predictions that can later be verified or refuted by observing the truth—for example, using past meteorological data to predict whether it will rain tomorrow. The "supervision" that guides our learning is the feedback we get tomorrow, when either it rains or it doesn't. And for much of the history of machine learning and statistical modeling, many applications, like this example, were focused on making predictions about nature or other large systems: predicting tomorrow's weather, predicting whether the stock market will go up or down (and by how much), predicting congestion on roadways during rush hour, and the like. Even when humans were part of the system being modeled, the emphasis was on predicting aggregate, collective behaviors.

The explosive growth of the consumer Internet beginning in the 1990s—and the colossal datasets it generated—also enabled a profound expansion of the ways machine learning could be applied. As users began to leave longer and longer digital trails— via their Google searches, Amazon purchases, Facebook friends and "likes," GPS coordinates, and countless other sources—massive datasets could now be compiled not just for large systems but also for specific people. Machine learning could now move from predictions about the collective to predictions about the individual.

And once predictions could be personalized, so could discrimination. While an abstract model of collective language use such as word2vec can be demonstrably sexist, it's hard to claim any particular woman has been hurt by it more than any other. When machine learning gets personal, however, mistakes of prediction can cause real harms to specific individuals. The arenas where machine learning is widely used to make decisions about particular people range from the seemingly mundane (what ads you are shown on Google, what shows you might enjoy on Netflix) to the highly consequential (whether your mortgage application is approved, whether you get into the colleges you applied to, what criminal sentence you receive). And as we shall see, anywhere machine learning is applied, the potential for discrimination and bias is very real—not in spite of the underlying scientific methodology but often because of it. Addressing this concern will require modifying the science and algorithms, which will come with its own costs.

YOU ARE YOUR VECTOR

To delve deeper into algorithmic notions of fairness, let's consider in more detail the standard framework for supervised learning, but in settings in which the data points correspond to information about specific, individual human beings. What this information contains will depend on the decisions or predictions we want our learned model to make, but it is typical to view that information as being summarized in a list x (technically, a vector) of properties (sometimes called attributes or features) deemed relevant to the task. For instance, if we are trying to predict whether applicants to a college will succeed if admitted (for some concrete, verifiable definition of success, like graduating within five years with at least a 3.0 GPA), the vector x for applicant Kate might include things such as her high school GPA, her SAT or ACT scores, how many extracurricular activities she lists on her application, a score given by an admissions officer to her application essay, and so on. If instead we are trying to decide whether to give Kate a loan for college, we might include all of the information above (since her success in college may impact her ability or willingness to repay the loan) as well as information about her parents' income, credit, and employment history. In either case, the goal of supervised learning is to build a model that makes a prediction y (like whether Kate will succeed in college) on the basis of the vector x (here summarizing Kate's relevant information) using "historical" data of the same $<x,y>$ form—in this case, past applicants to the college and whether they succeeded or not. (We'll return later to the important fact that the college really only learns the outcome y for those students it admitted in the past and not for those it rejected, so the college's own decisions are influencing and potentially biasing the data it collects.)

Note that Kate's parents' financial status isn't really information about Kate but still seems relevant to the loan prediction task, especially if Kate's parents will be cosigners on the loan. Looking ahead a bit, many debates about fairness in machine learning revolve in some way around what information "should" be allowed to be used in a given prediction task. Perhaps the most long-standing and contentious debate is over the inclusion of attributes such as race, gender, and age—properties that people cannot easily change, and that may seem irrelevant to the prediction task at hand, so that it may feel unfair to let them be used to make important decisions.

But what if it turns out that using race information about applicants really does result, on average, in more accurate predictions of collegiate success or loan repayment? Furthermore, what if those more accurate predictions result in some particular racial minority being discriminated against, in the sense that, all else being equal, members of that racial minority are less likely to be admitted to the college than others? Conversely, what if using race allows us to build accurate models that protect a group we wish to protect? Should we allow the use of race in these cases?

These questions do not have easy answers, and human judgments and norms will always need to play a central role in the debate. But the questions can be cast and studied scientifically—and even algorithmically—in a way that is certain to be central to the debate as well.

FORBIDDEN INPUTS

The question of what types of information should be allowed in making various decisions about people has been around for a long time, and is even the basis for significant bodies of law (for instance, in lending decisions and credit scoring, the direct use of race is generally illegal). But it has acquired greater urgency in the Internet era, where so much more data is gathered about people and is available for algorithmic decision-making—whether we know it or not. It is therefore tempting to assert that we can solve fairness problems simply by refusing to allow models to have access to things such as racial or gender data if we deem these irrelevant to the task at hand. But as we discuss elsewhere, it is difficult to confidently assert that any information about you is "irrelevant" in making almost any decision, because of the very strong correlations between attributes—so removing these features often really will diminish accuracy. Worse, removing explicit reference to features such as race and gender won't be enough to guarantee that the resulting model doesn't exhibit some form of racial or gender bias; in fact, as we will see, sometimes removing racial features can even exacerbate the racial bias of the final learned model.

If someone knows what kind of car you drive, what kind of computer and phone you own, and a few of your favorite apps and websites, they might already to be able to make a pretty accurate prediction of your gender, your race, your income, your political party, and many other more subtle things about you. More simply, in many areas of the United States your zip code is unfortunately already a pretty good indicator of your race. So if there is some property P of people that is, on average, relevant or informative in predicting whether they will repay loans, and apparently irrelevant properties Q, R, and S can be combined to accurately predict property P, then in fact Q, R, and S are not irrelevant at all. Moreover, removing property P from the data won't remove the algorithm's ability to make decisions based on P, because it can learn to deduce P from Q, R, and S. And given that these combinations may involve many more than just a few properties, and may be so complicated as to be beyond human understanding (yet not beyond algorithmic discovery), defining fairness by forbidding the use of certain information has become an infeasible approach in the machine learning era. No matter what things we prefer (or demand) that algorithms ignore in making their decisions, there will always be ways of skirting those preferences by finding and using proxies for the forbidden information.

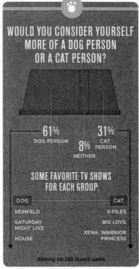

Fig. 9. Illustration of correlations between seemingly unrelated human attributes, such as between preferences for dogs or cats and favorite television shows. From data collected by the former tech startup Hunch (later acquired by eBay).

In other words, it has become virtually impossible to enforce notions of fairness that work by trying to restrict the inputs given to a machine learning or algorithmic decision-making process—there are just too many ways of deliberately or inadvertently gaming such efforts due to the number and complexity of possible inputs. An alternative approach is to instead define fairness relative to the actual decisions or predictions a model makes—in other words, to define the fairness of the model's outputs y rather than its inputs x.

While we will see that this approach has been more successful, it also is not without its own drawbacks and complexities. In particular, it turns out that there is more than one reasonable way of defining fairness for the predictions made, and these different ways can be in conflict with each other—so there's no way to have it all. And even if we settle on just one of them, the predictions made by a model obeying a fairness constraint will, as a general rule, always be less accurate than the predictions made by a model that doesn't have to; the only question is how much less accurate.

In other words, these kinds of more qualitative decisions and judgments—which type of fairness notion to use, when the reduction in accuracy is worth the gain in fairness, and many others—must remain firmly in the domain of human decision-making. The science part can begin only once society makes these difficult choices. Science can shed light on the pros and cons of different definitions, as we'll see, but it can't decide on right and wrong.

DEFINING FAIRNESS

The simplest notion of fairness applied to the predictions or decisions of a model is known as statistical parity. Like many definitions of fairness, defining statistical parity requires that we first identify what group of individuals we wish to protect. For concreteness, let's imagine a planet just like Earth except that there are only two races of people, Circles and Squares. Suppose for some reason we are concerned about discrimination against Squares in the granting of loans by a lender, so we ask that race be a protected attribute. Statistical parity simply asks that the fraction of Square applicants that are granted loans be approximately the same as the fraction of Circle applicants that are granted loans. That's all. The definition doesn't specify how many loans we have to give, or which particular Circle and Square citizens should receive them—it's just a crude constraint saying that the rate of granted loans has to be roughly the same for both races. Note that while our concern might have been discrimination against Squares, the definition is two-sided and thus also demands we not discriminate against Circles (though we could define a one-sided variant if we wanted).

Statistical parity is certainly some form of fairness, but generally a weak and flawed one. First, let's recall the framework of supervised learning, where loan applicants like Kate have specific individual properties summarized in their vector x, and there is some "true" outcome y indicating whether they will repay a loan. Statistical parity makes no mention of x at all—we could satisfy statistical parity by ignoring x entirely and picking an entirely random 25 percent of Circles and Squares to give loans to! This seems like a very poor algorithm for lending decisions, because we are entirely blind to the properties of individuals.

However, if we think about it a bit more carefully, this objection isn't that serious once we realize that statistical parity doesn't specify the goal of the predictions our model makes but is simply a constraint on those predictions. So the fact that there is a bad algorithm—random lending—that perfectly satisfies statistical parity does not mean there are not also good algorithms, ones that give loans to the "right" Circles and Squares. The goal of the algorithm might still be to minimize its prediction error, or maximize its profits; it's just that now it has to work toward this goal while being constrained to give out loans at equal rates. And in some sense it's reassuring that random lending obeys statistical parity, because it makes it immediately clear that the definition can be achieved somehow—which is not true of all fairness definitions.

Moreover, sometimes random lending might actually be a *good* idea, since it lets us obey statistical parity while we gather data. If we are a new lender and know nothing about the relationship between applicant attributes and loan repayment (i.e., between the x's and the y's), we can give out random loans for a while until we have enough <x,y> pairs to make more informed decisions, while still being fair (according to statistical parity) in the meantime. In machine learning, this would be called *exploration*—a period in which we are focused not on making optimal decisions but on collecting data. There might also be settings where the deliberate blindness of random decisions is desirable for its own sake—for instance, in distributing a limited number of free tickets to a public concert, where we don't view some candidates as more deserving or qualified than others.

A second and more serious objection is that statistical parity also makes no mention of the y's, which here represent the ultimate creditworthiness of each applicant. In particular, suppose it is the case that for some reason (and there might be many) Squares in the aggregate really are worse lending risks overall than Circles—for example, suppose that 30 percent of Circle applicants will repay loans, but only 15 percent of Square applicants will. (See Figure 10.) If we managed to find a perfect predictive model—that is, one that takes the x of any applicant, Circle or Square, and always correctly predicts whether that individual will repay the loan or not—then statistical parity forces us to make some difficult choices, because the actual repayment rates of the two races are different, but fairness requires us to grant loans at the same rates.

For instance, we could obey statistical parity by granting loans to exactly the 15 percent of Square applicants who will repay and to half of the 30 percent of Circle applicants who will repay. But this might also feel unfair—especially to the other 15 percent of creditworthy Circle applicants to whom we unjustly deny loans. And if we make money by giving loans to people who will repay them and lose money by giving loans to those who won't, then the lender is also making less money than it could. On the other hand, we could also obey statistical parity by giving loans to all 30 percent of repaying Circle applicants, but then we'd have to also give loans not just to the 15 percent of repaying Square applicants but also to another 15 percent of defaulting Square applicants in order to equalize the loan rates for the two populations. And now we'd lose money on these.

In other words (and again in the language of machine learning), while statistical parity is not at odds with exploration, it is at odds with *exploitation*—that is, with making optimal decisions—any time the optimal thing to do from an accuracy perspective differs between the two populations. In such cases, we can't simply optimize our

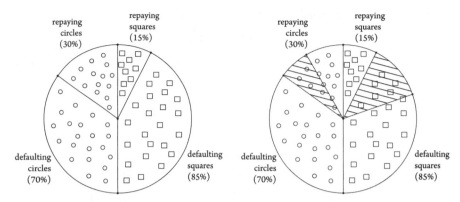

Fig. 10. Illustration of the tension between statistical parity and optimal decision-making. To obey statistical parity the lender must either deny loans to some repaying Circle applicants (shaded) or give loans to some defaulting Square applicants (shaded).

accuracy; we can only try to maximize it subject to the constraint of statistical parity. This is what the two solutions above do, in different ways—one by denying loans to creditworthy Circle applicants and the other by granting loans to Square applicants we know (or at least predict) will default. And while we shall next see that there are various kinds of improvements we can make to this coarse fairness constraint, the tension between fairness and accuracy will always remain, but can be made quantitative. In the era of data and machine learning, society will have to accept, and make decisions about, trade-offs between how fair models are and how accurate they are.

In fact, such trade-offs have always been implicitly present in human decision-making; the data-centric, algorithmic era has just brought them to the fore and encouraged us to reason about them more precisely.

ACCOUNTING FOR "MERIT"

The problem with statistical parity—that it can be violated even when making "perfect" decisions, if the Circles and Squares differ in their creditworthiness—can be remedied by requiring that we evenly distribute the *mistakes* we make, rather than evenly distributing the loans we give.

In particular, we could ask that the rate of false rejections—decisions by the model to deny a loan to an applicant who would have repaid—be roughly the same for the Circle and Square populations. Why should this be considered a notion of fairness? If we view the creditworthy individuals who are denied loans as being the ones harmed, this constraint requires that a random creditworthy Circle applicant and a random creditworthy Square applicant have the same probability of being harmed. In other words, all else being equal, your race doesn't affect the probability with which our algorithm will harm you. And if we could somehow never make any mistakes at all—achieving perfect accuracy—that would be deemed fair by this notion, since then the rates of false rejections would be zero for both populations, and thus equal.

But now it's also "fair" if our model mistakenly rejects, say, 20 percent of the population of Square applicants who would repay their loans … as long as it also mistakenly rejects 20 percent of the population of Circle applicants who would repay their loans. So here we are evenly distributing not the loans themselves (as in statistical parity), but rather the mistakes we make in the form of false rejections. This opens the door to building predictive models that are imperfect (an inevitability in machine learning, as discussed shortly) while still being fair according to this new definition, which is naturally called equality of false negatives. (We can just as easily define equality of false positives for settings in which such mistakes represent the greater harm.)

Of course, if you are one of the creditworthy Square applicants who was rejected for a loan, this might still seem unfair, and it might not comfort you to know that your own unjust treatment is being balanced by similar injustices to creditworthy Circles.[1] That's because both statistical parity and equality of false negatives are providing protections for groups (in this case the two races), but not for specific individuals in those groups, a topic we shall return to a bit later.

Since perfect decision-making is now deemed fair, we might be tempted to think that equality of false negatives eliminates the tension between fairness and accuracy. Unfortunately, while it does so in theory—namely, if from applications x we really could perfectly predict repayments y—in practice it does not, due to the cold realities of machine learning.

Those realities include the fact that the real world is messy and complicated, and even ignoring fairness entirely, it's rare to find a machine learning problem where there is sufficient data and computational power to find a model that makes perfect predictions—if such a model is even possible in principle, given that we can't hope to measure absolutely every salient property of loan applicants. And once we admit that our models will inevitably be imperfect, it's easy to find both cartoon and real examples of the tensions between accuracy and equality of false negatives.

FAIRNESS FIGHTING ACCURACY

As an example, let's consider a simplistic problem in which we have to decide whom to admit to the fictional St. Fairness College based solely on SAT scores. Again, members of the majority Circle and Square populations are applying, and it turns out that the majority of applicants are Circles. Furthermore, Circle applicants tend to be wealthier and thus can afford SAT preparation courses and pay for multiple retakes

---

[1] Sydney Morgenbesser, who was a law professor at Columbia University, was reputed to have said about the aftermath of campus protests in 1968, that the police assaulted him unjustly, but not unfairly. Asked to explain, he said that "They beat me up unjustly, but since they did the same thing to everyone else, it was not unfair."

of the exam. The Square applicants are less wealthy and generally take the exam once, with less practice and preparation. Not surprisingly, the SAT scores of the Circles are on average higher than for the Squares, but for superficial reasons. It turns out both populations are equally well prepared for collegiate success. In particular, suppose that the percentage of Circles who would succeed if admitted to St. Fairness is equal to the percentage of Squares who would succeed; it's just that the Circles have inflated SAT scores.

If our model is a simple threshold rule—we admit any applicant whose SAT score is above some cutoff—then we simply may not be able to make perfect predictions, and furthermore, even choosing the most accurate model may badly violate fairness. The fundamental problem is that since the Circles are the majority class, the most accurate model— that is, the one that minimizes the number of mistakes it makes in the aggregate—will set the threshold largely based on the SAT scores and collegiate success of the Circles, since by definition the rate of mistakes on the majority group counts more toward aggregate error than the rate of mistakes on the minority group. This comes at the expense of discrimination (a higher false rejection rate) against the Squares.

To illustrate this, suppose our historical applicant dataset looks like the following:

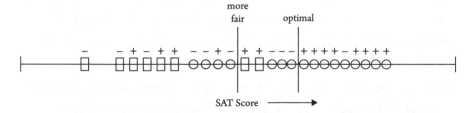

In this figure, circles represent Circle applicants and squares represent Square applicants. The position along the line indicates SAT scores, with higher scores to the right. A "+" symbol above an applicant indicates this person succeeded at St. Fairness, while a "−" indicates he or she did not.

On this data, the best model from a pure accuracy perspective is the cutoff labeled as "optimal." If we admit only applicants with this SAT score or higher, we make exactly seven mistakes: the one Circle − above the cutoff, the one Circle + below the cutoff, and the five Square +'s below the cutoff. But this means we would have falsely rejected these five successful Square applicants, whereas we would have falsely rejected only one Circle applicant. This violates the equality of false negatives notion of fairness. And if we use this cutoff to make decisions about future applicants, we should generally expect the disparity to be at least as bad as it was on the historical training data.

Of course, other models are possible. Moving the cutoff lower—for instance, to the line labeled "more fair"—improves fairness, according to equality of false negatives metric, by accepting two additional successful Squares, but it worsens accuracy, as we now make a total of eight mistakes instead of seven. The reader can confirm that even on this simple dataset, improving fairness will degrade accuracy, and vice versa.

Let's examine a couple of objections to this example. The first is that it is indeed a cartoon—no college would base admissions solely on SAT scores, but would instead build a more complex model incorporating many other factors. But in science generally, and in algorithm design specifically, if bad things can already happen in simple examples, they will also tend to happen in more complex ones—perhaps to an even greater extent. And the recent empirical machine learning literature is rife with examples of real-world problems in which building the optimal model for predictive accuracy results in demonstrable unfairness to some subpopulation. So increased complexity is not coming to our rescue on this issue.

A second objection is that the problem arises from our model not accounting for the fact that Circle and Square SAT scores differ for superficial reasons unrelated to collegiate success. If we know or detect statistically that the distributions of Circle and Square scores are different, why not build separate models for the two populations? For instance, in Figure 12 we show the same data with separate

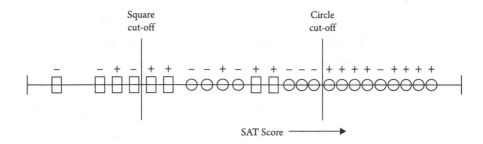

cutoffs for Circles and Squares. This hybrid model would make only three mistakes (two for the Circle cutoff and 1 for the Square), better than the seven mistakes for the aggregate single cutoff previously discussed, and it is also more fair, since it falsely rejects the same number (one) of Circle and Square applicants. So we've permitted a more complex model but have improved on both criteria.

This might indeed be a good thing to do, increasing both fairness and accuracy at the same time. In fact, it's not so different from how certain affirmative action policies are implemented. But note that this model (which really involves first picking which submodel to use based on the applicant's race) now explicitly uses race as an input, which, as we discussed above, is something some notions of fairness (and many laws) forbid, since race can equally well be used to increase rather than reduce discrimination. If we removed race as an input, it wouldn't be possible to implement this hybrid model. Even if we can use a model like this, it doesn't necessarily solve the problem—what if SAT scores are more predictive of college success for one population than another? For example, suppose the optimal predictive model for the Circle population alone has a false rejection rate of 17 percent and the optimal predictive model for the Square population alone has a false rejection rate of 26 percent. We're still discriminating against the Square population according to equality of false negatives, even if it might be less than with a single, common, race-blind model.

Note that the gender bias observed in the word embeddings model we discussed at the start of the chapter can be blamed on latent human bias that was present in the data. The algorithm simply picked up on the ways in which human beings used language—on reflection, how could we have expected it to do otherwise? But in the lending and admissions prediction problems we have just discussed, we can't as easily blame human bias in the data. We have assumed that the labels in our data are correct—anyone labeled in our dataset as able to succeed in college really is, and vice versa (of course, things only get worse if the labels might be wrong as well). The disparity in false rejections that emerges from our final admissions algorithm is the natural result of an algorithm trying to optimize for predictive accuracy—an emergent phenomenon that can't be blamed only on the class of models, the objective function, or any part of the data. It's just that when maximizing accuracy across multiple different populations, an algorithm will naturally optimize better for the majority population, at the expense of the minority population—since by definition there are more people from the majority group, and hence they contribute more to the overall accuracy of the model.

So there is simply no escaping that predictive accuracy and notions of fairness (and privacy, transparency, and many other social objectives) are simply different criteria, and that optimizing for one of them may force us to do worse than we could have on the other. This is a fact of life in machine learning. The only sensible response to this fact—from a scientific, regulatory, legal, or moral perspective—is to acknowledge it and to try to directly measure and manage the trade-offs between accuracy and fairness.

NO SUCH THING AS A FAIR LUNCH

How might we go about exploring this trade-off in a quantitative and systematic fashion—in other words, algorithmically? From the introduction—where we described the process of gradually adjusting a line or curve separating positive points from negative points, as well as the fancier but similar backpropagation algorithm for neural networks—we already have a sense of how machine learning goes about maximizing predictive accuracy on a dataset in the absence of any fairness constraint. On our St. Fairness College dataset, this process would entail searching through the possible values of the SAT cutoff for the one that minimized the total number of mistakes made (successful students rejected and unsuccessful students accepted, ignoring race). So even though the algorithmic details can be complicated for rich model classes, the basic idea is just a search for the model with the lowest overall error.

But we could equally well search for the model that instead minimized the overall unfairness. After all, for any proposed SAT cutoff, we can easily compute its "unfairness score" by just taking the magnitude of the difference between the number of falsely rejected Circle students and the number of falsely rejected Square students. Using the same principles as for standard error-minimizing machine learning, we could instead design algorithms for unfairness-minimizing machine learning.

Better yet, we could consider both criteria simultaneously. With each model we now associate two numbers: the number of mistakes it makes on the data and its unfairness score on the data. If we had an algorithm that could enumerate these numbers for all models under consideration, we could try to pick the one yielding the "best" trade-off. But what do we mean by best? Consider our dataset again, and the two race-blind cutoffs we first examined:

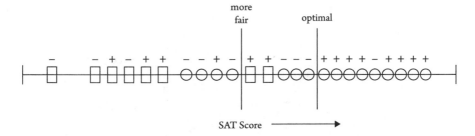

Which is better—the "optimal" cutoff, which makes seven mistakes and has an unfairness score of 4, or the "more fair" cutoff, which makes eight mistakes and has an unfairness score of 2? There is no universally

right answer, because each of these models is better on one criterion and worse on the other. They are thus incomparable, and we
should consider both to be reasonable candidates.

But sometimes there are models that really are worse. Consider the
cutoff obtained by moving the error-optimal cutoff line three circles to
its left, thus now accepting those three unsuccessful students. This model
now makes ten mistakes and has the same unfairness score (4) as the
optimal cutoff. So it ties the optimal cutoff on unfairness but has strictly
higher error. There's no plausible circumstance in which we'd prefer this
new model to the optimal one, because we can improve one of our objectives without paying for it in the other—in the language of machine
learning, the new model is *dominated* by the error-optimal one. In contrast, neither the "optimal" or "more fair" models dominates the other.

We can generalize this idea across our entire model space. Suppose
we enumerated all the possible numerical pairs *<error, unfairness>*
achieved by the models we are considering (e.g., SAT cutoffs).
Schematically, these pairs would give us a cloud of points that might
look something like the following:

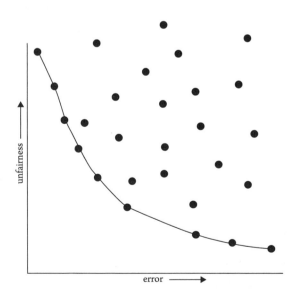

So each point corresponds to a different model; the x-coordinate of the point is the model's error, and the y-coordinate is its unfairness score. (For instance, the optimal cutoff would be a point at $x = 7$ and $y = 4$.) Here we have drawn a curve connecting the set of undominated models, which form the southwest (down and to the left) boundary of the set of points. The key thing to realize is that any model that is *not* on this boundary is a "bad" model that we should eliminate from consideration, because we can always improve on either its fairness score or its accuracy (or both) without hurting the other measure by moving to a point on this boundary.

The technical name for this boundary is the Pareto frontier or Pareto curve, and it constitutes the set of "reasonable" choices for the trade-off between accuracy and fairness. Pareto frontiers, which are named after the 19th-century Italian economist Vilfredo Pareto, are actually more general than just accuracy-fairness trade-offs, and can be used to quantify the "good" solutions to any optimization problem in which there are multiple competing criteria. One of the most common examples is the "efficient frontier" in portfolio management, which quantifies the trade-off between returns and risk (or volatility) in stock investing.

The Pareto frontier of accuracy and fairness is necessarily silent about which point we should choose along the frontier, because that is a matter of judgment about the relative importance of accuracy and fairness. The Pareto frontier makes our problem as quantitative as possible, but no more so.

The good news is that generally speaking, whenever we have practical algorithms for "standard," accuracy-only machine learning for a class of models, we also have practical algorithms for tracing out this Pareto frontier. These algorithms will be a little bit more complicated— after all, they must identify a collection of models rather than just a single one—but not by much. For instance, one algorithmic approach is to invent a new, single numerical objective that takes a weighted combination of error and the unfairness score. Thus we might ascribe a "penalty" to a model that looks like 1/2 times its error plus 1/2 times its unfairness, so the error-optimal cutoff would evaluate to $(1/2)7 + (1/2)4 = 5\ 1/2$. We then find the model that minimizes this new weighted penalty, which equally weights error and unfairness. It turns out that the model minimizing this weighted penalty must be one of the points on the Pareto frontier. If we then change the weightings— say, to 1/4 times error plus 3/4 times the unfairness score—we will find another point on the Pareto frontier. So by exploring different combinations of our two objectives, we "reduce" our problem to the single-objective case and can trace out the entire frontier.

While the idea of considering cold, quantitative trade-offs between accuracy and fairness might make you uncomfortable, the point is that there is simply no escaping the Pareto frontier. Machine learning engineers and policymakers alike can be ignorant of it or refuse to look at it. But once we pick a decision-making model (which might in fact be a human decision-maker), there are only two possibilities. Either that model is not on the Pareto frontier, in which case it's a "bad" model (since it could be improved in at least one measure without harm in the other), or it is on the frontier, in which case it implicitly commits to a numerical weighting of the relative importance of error and unfairness. Thinking about fairness in less quantitative ways does nothing to change these realities—it only obscures them.

Making the trade-off between accuracy and fairness quantitative does *not* remove the importance of human judgment, policy, and ethics—it simply focuses them where they are most crucial and useful, which is in deciding exactly which model on the Pareto frontier is best (in addition to choosing the notion of fairness in the first place, and which group or groups merit protection under it, both of which we discuss shortly). Such decisions should be informed by many factors that cannot be made quantitative, including what the societal goal of protecting a particular group is and what is at stake. Most of us would agree that while both racial bias in the ads users are shown online and racial bias in lending decisions are undesirable,

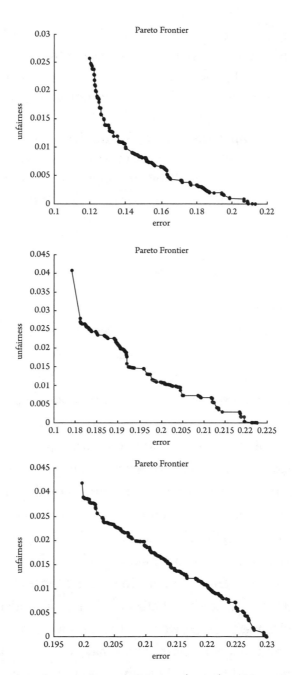

Fig. 15. Examples of Pareto frontiers of error (x axis) and an unfairness measure (y axis) for three different real datasets. The curves differ in their shapes and the actual numeric values on the error and fairness axes, thus presenting different trade-offs.

the potential harms to individuals in the latter far exceed those in the former. So in choosing a point on the Pareto frontier for a lending algorithm, we might prefer to err strongly on the side of fairness—for example, insisting that the false rejection rate across different racial groups be very nearly equal, even at the cost of reducing bank profits. We'll make more mistakes this way—both false rejections of creditworthy applicants and loans granted to parties who will default—but those mistakes will not be disproportionately concentrated in any one racial group. This is the bargain we must accept for strong fairness guarantees.

FAIRNESS FIGHTING FAIRNESS

Even before we arrive at the role of human judgment in the choice of a model on the Pareto frontier, there is the question of which fairness notion we want to use in the first place. As we've already seen, there is more than one reasonable choice. Statistical parity might be appropriate in settings where we simply want to distribute some opportunity, like free tickets to a concert, equally across groups, and there is no notion of merit (like creditworthiness) that is relevant. Approximate equality of false negatives (rejections) across groups might be appropriate in lending decisions. In picking whose tax returns to audit, equality of false positives (audits that discover nothing illegal) across groups might be the goal, since here it is the false positives—law-abiding citizens who are nevertheless subjected to a costly audit—who are harmed. And there are other reasonable fairness definitions where these came from.

In the same way that it was natural to hope for models that are as accurate and fair as possible, we might also hope to have it all when it comes to definitions of fairness. There may always be a trade-off between accuracy and fairness, but why not at least have our fairness notion be as strong as possible? For example, why not define fairness to mean satisfying statistical parity *and* equality of false negatives *and* equality of false positives *and* whatever else we can think of?

Alas, as with the Pareto frontier, we again encounter some stark barriers to all-encompassing definitions of fairness. It turns out there are certain combinations of fairness criteria that—although they are each individually reasonable—simply cannot be achieved simultaneously, even if we ignore accuracy considerations. There are mathematical theorems demonstrating this impossibility. One example is the combination of equality of both false positive and false negative rates across groups, along with another fairness notion called equality of positive predictive value. This simply asks that (for example) among those people the algorithm recommends be granted a loan, the repayment rates across racial groups are roughly the same. This is a desirable property for a predictive algorithm to have, because if the algorithm doesn't have it—that is, if the Circle applicants it recommends for loans end up making the bank less money than the Square applicants—the human decision-makers ultimately responsible for making loans will have a strong incentive to let race influence their decisions when they decide whether or not to follow the recommendations of their model. And they will be rational to do so, since if the model doesn't have equal positive predictive value for both races, its predictions for Circle applicants really will mean something different than its predictions for Square applicants.

So here we have a situation in which there are three different fairness definitions, each of which is entirely reasonable and desirable, and each of which can be achieved in isolation (albeit at some cost to accuracy)—but which together are simply impossible to achieve. Thus in addition to trade-offs with accuracy for any single fairness definition, there are even trade-offs we must make between fairness notions.[2]

---

[2] This is an interesting contrast to what we saw in the privacy chapter, where there does seem to be a single framework (differential privacy) that captures much of what we could (reasonably) want in a privacy definition and yet still permitted powerful uses of data such as machine learning. In other words, we already know that the study of algorithmic fairness will necessarily be "messier" than the study of algorithmic privacy and that we will have to entertain multiple, incompatible

These stark mathematical constraints on fairness are somewhat depressing, but they also identify and reinforce the central role that people and society will always have to play in fair decision-making, regardless of the extent to which algorithms and machine learning are adopted. They reveal that while algorithms can excel at computing the Pareto frontier once we commit to a definition of fairness, they simply cannot tell us which definition of fairness to use, or which model on the frontier to choose. These are subjective, normative decisions that cannot (fruitfully) be made scientific, and in many ways they are the most important decisions in the overall process we have been describing.

PREVENTING "FAIRNESS GERRYMANDERING"

There's one more crucial and subjective decision in this process that we've ignored so far, and that's the choice of which groups of individuals we want to protect in the first place. We've given examples of concerns over gender bias in word embeddings and racial discrimination in lending and college admissions. But race and gender are just two of many attributes we might want to consider. Extensive debate has also taken place over discrimination based on age, disability status, wealth, nationality, sexual orientation, and many other factors. The US Equal Employment Opportunity Commission even has regulations forbidding discrimination based on any type of "genetic information," an extremely broad category that clearly includes race and gender but also much more, including genetic factors yet to be discovered. As with the choices of fairness definition or which model on the resulting Pareto frontier we want to use, there's no "right answer," and no sensible role for algorithms or machine learning, in the choice

definitions of fairness. We might wish things were otherwise, but we must nevertheless proceed. But perhaps it does increase our appreciation of the strengths of differential privacy.

of which attributes or groups of people we want to protect—this is a decision for society to make.

One theme running throughout this book is that algorithms generally, and especially machine learning algorithms, are good at optimizing what you ask them to optimize, but they cannot be counted on to do things you'd like them to do but didn't ask for, nor to avoid doing things you don't want but didn't tell them not to do. Thus if we ask for accuracy but don't mention fairness, we won't get fairness. If we ask for one kind of fairness, we'll get that kind but not others. As we have seen, sometimes these tensions and trade-offs are mathematically inevitable, and sometimes they arise just because we didn't specify everything we did and didn't want.

This same theme holds true for the choice of which groups we protect. In particular, one recently discovered phenomenon is what we might call "fairness gerrymandering," in which multiple overlapping groups are protected, but at the expense of discrimination against some intersection of them. For example, imagine we want to distribute free tickets to see the Pope and want to protect both gender and race—so the same fraction of men and women should receive tickets, and also the same fraction of Circle and Square people (continuing our use of a fictitious two-race population). Suppose we have enough tickets for 20 percent of the population overall to see the Pope, and there are equal numbers of all four attribute combinations (Circle men, Circle women, Square men, Square women)—say twenty of each, for a total population of eighty, and thus we must distribute sixteen tickets total. (See Figure 16.)

We can probably agree that the "most fair" solution would be to give 4 tickets each to Circle men, Circle women, Square men, and Square women. But that's not what we ask for when we specify that we want to be fair with respect to race and gender *separately*. From that perspective, an equally fair solution is the "gerrymandered" one that gives eight tickets to Circle men, eight tickets to Square women, and no tickets to the other two groups. We still have eight tickets going to men, eight

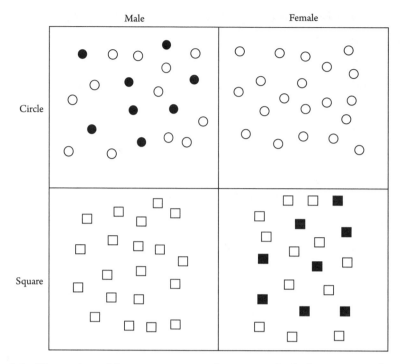

Fig. 16. Illustration of fairness gerrymandering, in which the winners of tickets (shaded circles and squares) are disproportionately concentrated in small subgroups despite being both gender- and race-fair separately.

to women, eight to Circles, and eight to Squares. It's just that we've concentrated those tickets in certain more refined subgroups at the expense of others (namely, Circle women and Square men).

One might ask who would come up with such a complex and discriminatory solution. The answer is that machine learning would, if it improves the accuracy of some prediction task—even by a tiny amount. We didn't ask that these more refined subgroups also be protected, only that the top-level attributes of race and gender be protected. If we also wanted such subgroup protections, we should have said so. And once we see this problem, there seems to be the potential for a kind of infinite regress, in which despite avoiding discrimination by race, gender, age, income, disability, and sexual orientation in isolation, we find ourselves with a model that, for example, unfairly treats disabled gay Hispanic women over age fifty-five making less than $50,000 annually.

While it is still early days for such topics, recent research has suggested improved algorithms for coping with fairness gerrymandering. These algorithms have the somewhat intuitive and appealing form of a two-player game between a Learner and a Regulator. The Learner is always trying to maximize predictive accuracy but is constrained to be fair to possibly complex subgroups identified by the Regulator (such as disabled gay Hispanic women over age fifty-five making less than $50,000 annually). A back-and-forth ensues, in which the Regulator finds new subgroups suffering discrimination under the Learner's model so far, and the Learner attempts to correct this discrimination while still preserving as much accuracy as possible. This process is guaranteed to quickly result in a model that is fair to all subgroups the Regulator is interested in protecting—even if the Regulator is potentially concerned with a great many groups, of the sort that result when we (for example) consider all possible ways of taking intersections of different categories of people. Experimental work suggests that this much stronger notion of subgroup fairness can be met while still providing usefully accurate models. In fact, Figure 15 earlier in the chapter, showing Pareto curves between error and unfairness on real datasets, is using this gerrymander-free fairness measure.

Of course, once we contemplate fairness for more refined subgroups in a population, it's hard not to take things to their logical conclusion, which would be promises of protections for individuals rather than just groups. After all, if we take a traditional statistical fairness notion such as equality of false rejection rates in lending, if you are one of the creditworthy Square applicants who has been denied a loan, how comforting is it to be told that there was also a creditworthy Circle applicant who was rejected to "compensate" for your mistreatment?

But if we go too far down the path toward individual fairness, other difficulties arise. In particular, if our model makes even a single mistake, then it can potentially be accused of unfairness toward that one individual, assuming it makes any loans at all. And anywhere we apply machine learning and statistical models to historical data, there are bound to be mistakes except in the most idealized settings. So we can ask for this sort of individual level of fairness, but if we do so naively, its applicability will be greatly constrained and its costs to accuracy are likely to be unpalatable; we're simply asking for too much. Finding reasonable ways to give meaningful alternative fairness guarantees to individuals is one of the most exciting areas of ongoing research.

BEFORE AND BEYOND ALGORITHMS

Our focus in this chapter has been on machine learning algorithms, the predictive and decision-making models they output, and the tensions between accuracy and fairness. But there are other places in the typical workflow of machine learning in which fairness concerns can arise, both before we get to algorithms and models and after we've deployed them.

Let's begin with the "before"—namely, with the data collected and fed to a machine learning algorithm in the first place. Throughout much of the chapter we have implicitly assumed that this data is itself correct and not already corrupted by human bias. Our main goal was to design algorithms that do not introduce discrimination as a side effect of the optimization of predictive accuracy. But what if the data itself has been gathered as part of an already discriminatory process? For example, maybe we wish to predict crime risk, but we don't have data on who commits crimes—we only have data on who was arrested. If police officers already exhibit racial bias in their arrest patterns, this will be reflected in the data.

For another example, returning to our hypothetical admissions scenario at St. Fairness College, suppose human admissions officers historically had a much better understanding of the Circle applicants who form the majority population than they did of the minority Square applicants—being more familiar with the high schools attended by Circles, understanding Circle essays and extracurricular activities better, and generally being more informed about the Circle population. Maybe the admissions officers weren't consciously biased against the Square applicants; they just know the Circle applicants better. It would not be at all surprising if these admissions officers were much better at accurately picking Circle applicants who will succeed at St. Fairness than they are at picking Square applicants who will succeed. Then, remembering that the college only learns about the success or failure of applicants it chooses to admit and not the ones it rejects, the obvious outcome is that the $<x,y>$ data generated by past admissions decisions will make Circle applicants look much better than the Square applicants in aggregate—not because they are better in reality but only because of the relative expertise of the admissions officers and the skewed sample of data that they produce.

And if it is the case that in our historical data Circles look better than Squares, there is absolutely no reason to hope that a machine learning algorithm—even one carefully applying all of the anti-discrimination methodology we have been discussing in this chapter—won't learn a model favoring Circle applicants over Square applicants. The problem is that even if we equalize the false rejection rates between Circle applicants and Square applicants on the training data, we will not do so on general populations of Circles and Squares, because the training data is not representative of those general populations. Here the problem is not the algorithm but the mismatch between the input to the algorithm and the real world, caused by the bias already embedded in the data. And this is something we simply cannot expect the algorithm itself to discover and correct. It's just the machine learning version of the computer science adage "Garbage in, garbage out." We might call this version "Bias in, bias out."

The problems can become even more insidious. Sometimes decisions made using biased data or algorithms are the basis for further data collection, forming a pernicious feedback loop that can amplify discrimination over time. An example of this phenomenon comes from the domain of "predictive policing," in which large metropolitan police departments use statistical models to forecast neighborhoods with higher crime rates, and then send larger forces of police officers there. The most popularly used algorithms are proprietary and secret, so there is debate about how these algorithms estimate crime rates, and concern that some police departments might be in part using arrest data. Of course, even if neighborhoods A and B have the same underlying rates of crime, if we send more police to A than to B, we naturally will discover more crime in A as well. If this data is then fed back into the next update of our model, we'll "confirm" that sending more police to A than to B was the right thing to do, and send even more next time. In this way, even small random fluctuations in observed crime rates can lead to self-fulfilling prophecies of enforcement that have no underlying basis in reality. And if the initial training data in this process has not just innocent random fluctuations but is in fact the result of historical crime rates measured during a biased period of police deployment (e.g., one that disproportionately policed minority neighborhoods), the amplification is all but guaranteed. It's another example of a mismatch between the data fed to an algorithm and the world that data is meant to represent, but now accelerated with a feedback loop.

As we have seen, the design of fair machine learning algorithms can be made scientific and is (at least in principle) easy to implement in practice. It would be necessary for companies, organizations, and engineers to be aware of this science, to take it seriously, and to want to incorporate it into their code. Even complex computer programs and systems are often built by relatively small teams, so the number of people who need to be educated and trained is manageable. But what about the problems arising from biased data collection?

Unfortunately, in many cases these problems are as much social as algorithmic, and are accordingly more difficult. When data collection for college admissions or predictive policing is done by large, distributed, and heterogeneous groups of people, each with their own unknown strengths, weaknesses, and biases, the challenge of imposing clean and principled practices can feel daunting. And in many cases, the scientific solutions suggested by machine learning—such as only training on data gathered during "exploration" phases in which (say) random applicants to St. Fairness are granted admission without anyone looking at their applications—are simply impractical to implement for policy, legal, or social reasons. So while there is now quite a bit of solid science around fairness, there's much more to do to understand how to better connect the narrow purview of an algorithm in isolation with the broader context in which it is embedded. There is much we don't know—but that also makes it an exciting time to be working in this area.

Games People Play
(With Algorithms)

THE DATING GAME

In 2013, a journalist named Amanda Lewis wrote an insightful article for *LA Weekly* about her experiences with a recently launched online dating app named Coffee Meets Bagel. One of the app's novelties was to apply the notion of economic scarcity to romantic matchmaking. Instead of encouraging users to indiscriminately spam potential dates with a barrage of online flirts, nudges, and winks, Coffee Meets Bagel limited users to a single, algorithmically proposed match or date each day, which they could accept or reject. Presumably the idea was to raise the value or demand for matches by artificially restricting the supply.

But Lewis went on to detail other "economic" side effects of the app that were perhaps less intentional, and less desirable—side effects that can be understood via game theory, the branch of economics that deals with strategic interactions between groups of self-interested individuals. Coffee Meets Bagel invited users to specify racial, religious, and other preferences in their matches, which the algorithm would then try to obey in selecting their daily proposals.

Lewis described how after not specifying any racial preferences (or, more precisely, indicating that she was willing to be matched with people from any of the site's listed racial groups), she began to receive daily matches exclusively with Asian men. The problem was that if there were even a slight imbalance in the number of women who accept matches with Asian men, and the number of Asian men, there would be an oversupply of Asian men in the app's user population. And since the matching algorithm obeys users' stated preferences, the necessary consequence was that women who did not explicitly *exclude* Asian men from their preferences would be matched with them frequently.

Given the preferences selected by the rest of the user population, Lewis's "best response" (a game theory term)—that is, her only choice if she wanted to be matched with men from other races too—was to modify her stated preferences to say that she was *unwilling* to be matched with Asian men. She reluctantly did so, even though this was not what she originally wanted. Of course, this only exacerbates the original oversupply problem, creating a feedback loop that encourages other users to do the same.

It seemed as though Lewis had been cornered into choosing between two undesirable alternatives—cornered by the stated preferences of other users, and by an algorithm that blindly and myopically obeyed those preferences for each user individually, without regard for the macroscopic consequences. At least from Lewis's perspective, the system was trapped in what a game theorist might call a "bad equilibrium." If all of the users of the app could have simultaneously coordinated to change their preferences, they might all have been happier with their resulting set of matches—but each of them individually was helpless to escape this bad outcome. It's a bit like a run on the banks in a financial crisis—even though it makes us all collectively worse off, it's still in your selfish interest to withdraw your money before it's too late.

WHEN PEOPLE ARE THE PROBLEM

There are some important similarities and differences between the dilemma Lewis found herself in on Coffee Meets Bagel and the problems of fairness and privacy considered in earlier chapters. In all three settings, algorithms play a central role—algorithms acting on, and often building predictive models from, people's data. But in algorithmic violations of fairness and privacy, it seemed reasonable to view the algorithm as the "perpetrator" and people as the "victims," at least to a first approximation. As we saw, machine learning algorithms optimizing solely for predictive accuracy may discriminate against racial or gender groups, while algorithms computing aggregate statistics or models from behavioral or medical data may leak compromising information about specific individuals. But the people themselves were not conspirators in these violations of social norms—indeed, they may not even be aware that their data is contributing to a credit scoring or disease prediction model, and may not interact with those models at all. And since the problems we identified were largely algorithmic, we could propose algorithmic solutions that were better behaved.

The Coffee Meets Bagel conundrum is more nuanced. We might argue that Lewis is also a victim of sorts—she recounts feelings of guilt when the algorithm forces her to declare what feel like racist preferences, just in order to avoid being always matched with a homogeneous group. It seems unfair in a way not dissimilar to algorithmic discrimination. But the key difference is that we can no longer place the blame exclusively or even largely on the algorithm alone—the other users, and their competing preferences, are complicit in Lewis's dilemma. After all, it wasn't the algorithm's fault that there were too many Asian men in the system relative to the population of women who reported a willingness to date them. The algorithm was simply trying to act as a mediator of sorts, attempting to satisfy each user's dating preferences in light of those of other users. We might even say that the algorithm was doing the most obvious and natural thing with the data it was given, and that the real problem was the data—the preferences themselves.

We'll eventually see that despite the complicity of users, we shouldn't let algorithms off the hook quite so easily, and that in many settings in which user preferences are centrally involved, there are still algorithmic techniques that can avoid the bad equilibrium in which Lewis became trapped. In particular, sometimes there might be multiple equilibria, and an algorithm might be able to choose, or nudge its users toward, a better one. In the case of Coffee Meets Bagel, maybe *everyone*'s preferences were like Amanda's—wanting only a diversity of matches— and everyone felt trapped into entering preferences that weren't quite truthful. Maybe a different algorithm could have done better and incentivized everyone to enter their real preferences. And in other settings we might prefer an algorithm that doesn't encourage or implement any equilibrium at all, but instead finds a solution that makes the overall "social welfare" higher. But unlike the fairness and privacy chapters, to discuss these algorithmic alternatives, we need to put the users, and their preferences, on center stage. And this in turn leads us to the powerful concepts and tools of game theory.

JUMP BALLS AND BOMBS

Many readers may have encountered a little game theory, owing in part to its generality and its ability to sometimes generate counterintuitive insights about everyday scenarios. Informally speaking, an equilibrium in game theory can be described as a situation in which all participants are acting in their own self-interest, given what everyone else is doing. The key aspect of the definition—which we'll make a bit more precise when it's called for—is the notion of selfish, unilateral stability it embodies. It is assumed that each "player" in the system (such as a user of Coffee Meets Bagel) will behave selfishly (for example, by setting or changing her dating preferences) to advance her own goals, in response to similarly selfish behavior by others, and without regard to the consequences for other players or the global outcome.

An equilibrium is thus a kind of selfish standoff, in which all players are optimizing their own situation simultaneously, and no one can improve their situation by themselves. Technically speaking, the underlying mathematical notion of equilibrium we refer to here is known as a Nash equilibrium, named for the Nobel Prize–winning mathematician and economist John Forbes Nash, who proved that such equilibria always exist under very general conditions. We'll shortly have reason to consider non-equilibrium solutions to game-theoretic interactions, as well as alternative notions of equilibrium that are more cooperative.

When equilibrium is described as a selfish standoff, it's not particularly surprising that sometimes equilibrium can be undesirable to any particular individual in the system (like Amanda Lewis), or even to the entire population. In the words of the late economist Thomas Schelling (another Nobel Prize winner), who applied equilibrium analysis to things as varied as housing choices, traffic jams, sending holiday greeting cards, and choosing a seat in an auditorium, "The body of a hanged man is in equilibrium, but nobody is going to insist the man is all right."

While the competitive, selfish nature of our equilibrium notion might seem a bit cynical or depressing—everyone is simply out for themselves, and optimizing their choices and behavior in light of everyone else's greedy behavior—it can also provide valuable clues to why and how things can sometimes go wrong in settings in which there are conflicting preferences (like racial preferences in a dating app). And it does not preclude cooperative behavior if there just so happens to be a solution in which cooperation is in everyone's self-interest. Closest to our own selfish interests, it turns out that sometimes game theory can not only describe what might go wrong at equilibrium but also can provide algorithmic prescriptions for making the outcome better.

For much of its long and storied history—depending on how one counts, the field dates to at least the 1930s—game theory trafficked primarily in the precise understanding of simple and highly stylized versions of real-world problems. These stylizations could often be described by small tables of numbers specifying the payoffs of just two players (and therefore their preferences, since it is assumed that a player will always prefer whatever offers their highest payoff, in light of the opponent's behavior). Classic examples include Rock-Paper-Scissors (useful in the real world as an alternative to jump balls in recreational basketball), where, for instance, choosing Rock yields payoff +1 against Scissors, which in turn receives payoff -1. The equilibrium turns out to be both players uniformly randomizing among their choices, playing Rock, Paper, and Scissors with probability 1/3 each. This is the only solution with the aforementioned unilateral stability property—if I uniformly randomize, your best response is to do so as well, and if you do anything else (such as playing Paper even slightly more often than the other two choices), I'll exploit that and punish you (by always playing Scissors). Some readers may be even more familiar with Prisoner's Dilemma, another simple game that has a disturbing equilibrium in which both players sabotage each other to their mutual harm, even though there is a cooperative non-equilibrium outcome in which they both benefit. As the story goes, two accomplices to a crime are captured and held in separate cells. They can either "cooperate" with their accomplice and admit to nothing or "defect" and admit to the crime and testify against their partner. If your partner defects and testifies against you, you get a long sentence. If your partner cooperates, you get only a short one. And if you defect, the prosecutor offers to shave a little bit off the sentence you would have otherwise received. The problem is that if I cooperate, you can do even better by sabotaging me and defecting, and vice versa. When we both defect, we each experience close to the worst possible outcome. But since mutual cooperation is not unilaterally stable, we drag each other into the sabotaging abyss of equilibrium, hence the "dilemma".

Despite the simplicity of such games, they have occasionally been applied to rather serious and high-stakes problems. During the Cold War, researchers at the RAND Corporation (a long-standing think tank for political and strategic consulting) and elsewhere used game-theoretic models to try to understand the possible outcomes of US-Soviet nuclear war and détente—efforts that were memorably if darkly lampooned in the 1964 Stanley Kubrick film *Dr. Strangelove*, which ends with the Prisoner's Dilemma–like nuclear annihilation of the world. But the lasting influence and scope of game theory (which has also been widely applied to evolutionary biology and many other fields far from its origins) bears testament to the value of deeply understanding a "toy" version of a complex problem. By distilling strategic tensions down to tables of numbers with maybe only a few rows and columns, game theorists could solve exactly for the equilibrium and try to understand its ramifications for the real problem—which was usually considerably more complicated, messy, and imprecise.

As we shall see, the technological revolution of the last two decades has considerably expanded the scope and applicability of game-theoretic reasoning, while at the same time challenging the field to tackle problems of unprecedented scale and complexity—problems involving sophisticated algorithms operating on rich datasets generated by thousands, millions, or sometimes billions of users. Reducing such problems to simple models of the Rock-Paper-Scissors or Prisoner's Dilemma variety is entirely infeasible and would throw away too much salient detail to be even remotely useful. The matchmaking equilibrium determined by the dating preferences of the users of Coffee Meets Bagel simply isn't something that can be computed by hand and understood with just a few numbers. It would itself require an algorithm to compute, which in an informal sense is exactly what the app provides.

To tackle such challenges, the new field of algorithmic game theory has emerged and developed rapidly. It blends ideas and methods from classic game theory and microeconomics with modern algorithm design, computational complexity, and machine learning, with the goal of developing efficient algorithmic solutions to complex strategic interactions between very large numbers of players. At a minimum, it aspires to broadly understand what might happen in systems like Coffee Meets Bagel. At its best, it is not only descriptive but also prescriptive—as in the fairness and privacy chapters, telling us how to design socially better algorithms, but now in settings in which the incentives and preferences of users, and how we will examine in act on them, must be taken into account. These are the topics we will examine in this chapter.

THE COMMUTING GAME

To illustrate how the scale and power of modern technology have made algorithmic game theory relevant, let's consider an activity that many people engage in every day but may never have thought about as a "game" before: driving a car. Suppose you live in a busy metropolitan area with congested roads, and each day you must drive from your house in the suburbs to your workplace downtown. There is a complex network of freeways, highways, streets, thoroughfares, and back roads you must navigate, and the number of plausible routes you could take might be very large indeed. For instance, maybe the most straightforward route is to get on the freeway at the entrance nearest your home, get off at the exit nearest your workplace, and drive on the main surface streets before and after the freeway. But maybe one segment of the freeway often has bumper-to-bumper traffic during your commute time, so sometimes it's better to get off earlier, take some back roads through a residential neighborhood, and rejoin the freeway later. And on any given day, transient conditions—a traffic accident, road construction or closures, a ball game—might render your usual route much slower than some other alternative.

If you think about it, on a moderately long commute in a busy city the number of distinct routes you might take or at least try over time could be in the dozens or even hundreds. Of course, these different routes might overlap to varying degrees—maybe many of them use the freeway, and if you live on a cul-de-sac, they will always start by getting off your street—but each route is a distinct path through your local network of roads. In game theory terminology, your "strategy space"—the possible actions you might choose—is much larger than in simple games like Rock-Paper-Scissors, where by definition you only have three actions available.

So you have a lot of choices; but what makes this a "game"? It's the fact that if you're like most commuters, your goal or objective is to minimize your driving time. But the driving time on each of your many possible routes depends not just on which one you choose but also on the choices of all the other commuters. How crowded each route is determines your driving time as much or more than the length of the roads, their traffic lights, speed limits, and other fixed aspects. The more drivers who choose a given road, the longer the driving time for all routes that use that road, making them less attractive to you. Similarly, the fewer drivers there are on a road, the more you might want to choose a route that uses it (as long as the other segments on the route aren't too busy).

The combination of your hundreds of possible routes with the choices made by the tens of thousands of other commuters presents you with a well-defined, if mind-boggling, optimization problem: pick the route with the lowest total driving time, given the choices of all the other drivers. This is your "best response" in the commuting game. And it's not at all unreasonable for us to assume you will at least try to act selfishly and choose your best response (just as Amanda Lewis begrudgingly did on Coffee Meets Bagel). Who wants to spend more time commuting than they need to?

Note that while the complexity of this game is much greater than something like Rock-Paper-Scissors, the fundamental commonality is that the payoff or cost of any individual player's choice of action depends on the action choices of *all* the players. There are important differences as well. In Rock-Paper-Scissors the two players have the same payoff structure, whereas if you and I live and work in different places, our cost structures will differ (even though we still both want to minimize driving time for our own commute). And if you and I commute at different times of day, we aren't really even in the same round of the game. But these differences don't alter the fundamental view of commuting as just another (albeit very complex) game. And this means that as with Rock-Paper-Scissors and Coffee Meets Bagel, it makes sense—from both qualitative and algorithmic perspectives—to discuss its equilibrium, whether it is "good" or "bad," and whether there might be a better outcome.

YOUR SELFISH WAZE

Commuting has been the game we have described ever since roads became sufficiently congested that the choices of other drivers affected your own. But for many years this formulation wasn't particularly relevant, because people really didn't have the ability to truly or even approximately optimize their route based on the current traffic. Commuting was a game, but people couldn't play it very well. This is where technology changed everything—and, as we shall see, not necessarily for the collective good.

The first challenge in playing the commuting game is informational. As older readers will recall, for decades you had to plan your daily commute by cobbling together radio and television traffic reports that were both incomplete (perhaps covering only major freeways, and providing little or no information about the vast majority of roads) and inaccurate (since the reports were only occasional, perhaps on the half hour, and therefore often stale). But even if one could magically always have perfect and current traffic data for every road, there is a second, algorithmic challenge, which is computing the fastest route between two points in a massive network of roadways, each annotated by its current driving time.

In a relatively short period, navigation apps such as Waze and Google Maps have effectively solved these problems. The algorithmic challenge was actually the easier one—there have long been fast, scalable algorithms for computing fastest routes (or "shortest paths," as they are called in computer science) from known traffic. A classical one is Dijkstra's algorithm, named for the Dutch computer scientist who described it in the late 1950s. Such algorithms in turn allowed the informational problem to be solved by crowdsourcing. Even though early navigation apps operated on traffic data not much better than in the pre-Internet days, they could still at least suggest plausible routes through a complex and perhaps unfamiliar city—a vast improvement over the era of dense and confusing fold-up maps in the glove compartment. And once users started adopting the apps and permitting (wittingly or not) their location data to be shared, the apps now had thousands of real-time traffic sensors right there on the roadways.

This crowdsourcing was the true game-changer. Whatever pride you might have had in your navigational wizardry in your home city, the utility of a tool that automatically optimized your driving time in response to real-time, highly accurate, and granular traffic data on virtually every roadway anywhere was just too alluring to decline. User populations grew to the hundreds of millions, further improving traffic data coverage and accuracy.

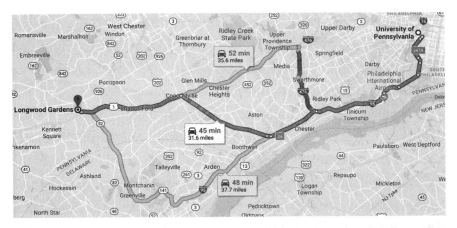

Fig. 17. Typical screenshot from Google Maps, showing just a few of the many hundreds or thousands of routes between two locations in the greater Philadelphia area. The suggested routes are ranked by lowest estimated driving time.

From our game-theoretic viewpoint, modern navigation apps finally allowed any player in the commuting game to compute her best response to all her "opponents" on the roads, anywhere and anytime. And there is little doubt that these apps are extraordinarily useful and efficient, and are doing the most obvious thing with the massive data at their disposal: looking out for the best interests of each individual user, finding their fastest route in light of the current traffic patterns.

THE MAXWELL SOLUTION

But there is another perspective worth considering, which is that because the apps are computing best responses for every player individually, they are driving the collective behavior toward the kind of competitive equilibrium we have discussed in Coffee Meets Bagel, Prisoner's Dilemma, and Rock-Paper-Scissors—the apps enable, and thus encourage, selfish behavior on everyone's part. And with Coffee Meets Bagel and Prisoner's Dilemma, we already have seen cases in which the resulting competitive equilibrium may not be something that any particular individual is happy about. Surely anyone with even moderate experience with city driving has encountered situations in which individually selfish behavior by everyone seems to make everyone worse off—for instance, in the jockeying and slithering that occur when merging down to a few lanes of traffic at the entrance to the Lincoln Tunnel in New York City.

What might be an alternative to individually selfish, collectively competitive driving? Surely no one believes we'd all be better off (at least in terms of driving time) if we rolled back the calendar and returned to the era of spotty traffic reports and folding maps. But now that we do have large-scale systems and apps with the ability to aggregate granular traffic data, compute and suggest routes to drivers, it might be worth considering making recommendations other than the obvious, selfish ones.

Let's consider a conceptually simple thought experiment. Imagine a new navigation app—we'll name it Maxwell, for reasons that will become clear later—that behaves similarly to Google Maps and Waze, at least at a high level. Like those apps, Maxwell gathers GPS and other location data from its users to create a detailed and up-to-date traffic map, and then for any user at any moment, it computes and suggests a driving route based on origin, destination, and the traffic. But Maxwell is going to use a very different algorithm to compute suggested routes— an algorithm with a different goal, and one that will lead to a different and better collective outcome than the competitive equilibrium.

Instead of always suggesting the selfish or best response route to each user in isolation, Maxwell gathers the planned origin and destination of every user in the system and uses them to compute a coordinated solution that is known in game theory as the maximum social welfare solution (hence the app's name, Maxwell). In the commuting game, the maximum social welfare solution is the one that minimizes the *average* driving time across the entire population, instead of trying to minimize the driving time of each user *individually* in response to the current traffic. By minimizing average driving time, Maxwell is maximizing the time people have to do other things, which is presumably a good thing.

It might seem like there shouldn't be any difference between these two solutions, but there is. A stylized but concrete example will be helpful here. Imagine that there is a large population of N drivers in a city, and all of them want to simultaneously travel from location A to location B. There are only two possible routes from A to B; let's call them the slow route and the fast route.

The slow route passes many schools, hospitals, libraries, restaurants, shops, and other places that generate a great deal of pedestrian traffic. It is littered with stop signs, crosswalks, speed bumps, and police making sure that all laws are obeyed. Because of this, it really doesn't matter how many drivers take the slow route. The real bottleneck is all the stop signs, crosswalks, speed bumps, and police. In other words, we are going to assume that the time it takes to travel from A to B on the slow route is independent of the number of drivers on it. To make things concrete, let's suppose that travel time is exactly one hour.

The fast route, on the other hand, is a freeway without speed limits or police, but it has limited capacity. If you're the only one driving on it, it can be very fast—almost instantaneous—to get from A to B. But the more drivers who take the fast route, the less fast it becomes. Specifically, let's assume that if M out of the N drivers take the fast route, the travel time for each of them is M/N hours. Since M is a whole number less than or equal to N, this means that the time it takes to travel the fast route is between $1/N$ (if only one driver takes

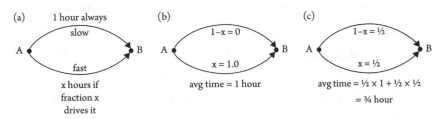

Fig. 18. Illustration of simple two-route navigation problem, with a fixed-driving-time slow route and a traffic-dependent fast route (a); equilibrium or Waze solution (b); Maxwell solution (c).

it), which is nearly zero if N is large, and N/N, which is one hour if everyone takes it. So in the worst case, the fast route isn't any faster than the slow route, but it depends on M. From your perspective as a driver, you'd like all the other $N - 1$ drivers to take the slow route, taking exactly an hour each, and for you to take the fast route and virtually teleport to the destination. Of course, none of the other drivers like your solution.

Now let's analyze the consequences of selfish behavior, of the kind enabled and even encouraged by existing navigation apps. If we think about it, such apps will recommend the fast route to the entire population of N drivers. This is because if the app recommended the slow route to even a small number of drivers—say, five—then these five drivers all experience the fixed one hour of slow route travel time, but any one of them would have been slightly better off taking the fast route, where the travel time will be $(N-5)/N = 1-5/N$ hours—just a shade less than an hour. So the competitive equilibrium that results from selfish routes is where everyone takes the fast route, which then becomes no faster than the slow route, and everyone's driving time is exactly one hour. Note that in this equilibrium, each individual driver is actually indifferent to which route is taken—the driving time for both is an hour—but if even one driver is on the slow route, the drivers on the fast route are strictly better off.

What is Maxwell going to do in the same situation? It is going to pick half of the drivers—let's say a random half—and suggest that they drive the slow route, and suggest the fast route to the other half. Before discussing why anyone would follow the suggestion to drive the slow route, let's analyze the average driving time in this alternative solution. Obviously the $N/2$ drivers taking the slow route will, as always, experience a driving time of one hour. The $N/2$ drivers taking the fast route will experience a driving time of only $(N/2)/N = 1/2$ hour each. So the average driving time across the entire population is $(1/2 \times 1) + (1/2 \times 1/2) = 3/4$ of an hour, or only forty-five minutes. It turns out this is the split of the population into the slow and fast route that minimizes the average driving time. (For readers who both took and remember some calculus, if we let x denote the fraction of the population on the fast route, the average driving time is simply $1 - x + x^2$, which is minimized at $x = 1/2$ and yields the 3/4 hour average.)

In other words, by suggesting routes with a different goal—one with an explicit concern for the collective benefit rather than individual self-interest—we can reduce the overall driving time significantly, by 25 percent in this case. And we can do so without making anyone worse off than they would have been in the competitive equilibrium. So there is a better alternative to the competitive equilibrium in our toy example, and the gains may generally be even greater in complex networks of roads in the real world.[1] (In 2018 a team of researchers from UC Berkeley presented empirical evidence that navigation apps indeed cause increased congestion and delays on side streets.) The question is whether and how we can actually realize this savings of collective driving time "in the wild."

MAXWELL'S EQUATIONS

The first challenge in implementing Maxwell is algorithmic. While it was a simple calculus exercise to find the socially optimal solution in our toy two-route example, how will Maxwell do it when confronted with colossal networks of real roads and freeways, and thousands or more drivers, all with different origins and destinations? At least the selfish routes suggested by Google Maps and Waze can be computed quickly on large-scale networks, using Dijkstra's algorithm.

Fortunately, it turns out that there are also fast, practical algorithms for computing the global solution that minimizes collective average

---

[1] A distinct but related side effect of selfish behavior in commuting is known as Braess's paradox, in which adding capacity to a network of roadways actually *increases* congestion (or closing roads decreases congestion), and which has been reported to have occurred in large cities such as Seoul, Stuttgart, and New York City. Such phenomena cannot occur under the Maxwell solution.

driving time in large networks, especially if the driving times on each road are a linear (i.e., proportional) function of the number or fraction of drivers on them (like the roads in our earlier example, or more realistic ones such as a road that hypothetically takes $1/4 + 2x$ hours to travel if a fraction x of the population drives on it). This proportional model actually seems like a reasonable one for real traffic, and we can easily envision deriving such models from the voluminous empirical data that services such as Waze already routinely collect, which provides samples of the driving times at different levels of traffic. And for such roads, the average driving time is then just a quadratic function (e.g., if a fraction x of drivers takes a $1/4 + 2x$ road, then the contribution to the overall average driving time from just this road will be $(1/4 + 2x)x = 1/4x + 2x^2$).

Even though Maxwell must solve a very high-dimensional problem—finding the exact fractions of drivers taking every road in the network, in a way that is consistent with everyone's origins and destinations and is socially optimal—it is a problem of a well-studied and well-understood type that has very practical algorithms. It is an instance of what are known as convex minimization problems, which can be solved by so-called gradient descent methods; this is just algorithm-speak for "walk downhill in the steepest direction to quickly get to the lowest point in the valley." In our context, this simply means that we start with an arbitrary assignment of driving routes and make incremental improvements to it until the collective driving time is minimized.

What if the driving times on the roads are not proportional to traffic but are more complex functions? For example, consider a hypothetical road whose driving time is $x/2$ for $x < 0.1$ but is $10x + 2$ for $x >\, = 0.1$. So the time it takes to drive this road takes a sudden, discontinuous jump once 10 percent or more of the population takes it. For more complex roads such as these, we do not know of fast algorithms that are always guaranteed to find the socially optimal solution, but we do know of good techniques that work well in practice. And in these more complex cases, the improvement of the socially optimal solution over the selfish equilibrium can be much greater than in the proportional-road setting. Thus at least the algorithmic challenges in implementing Maxwell seem surmountable.

CHEATING ON MAXWELL

But as is often the case in settings in which human preferences and game theory are involved, the biggest challenge Maxwell would face in the real world has less to do with good algorithms and more to do with incentives. Specifically, why would anyone ever follow the advice of an app that sometimes doesn't suggest the route that would be fastest for him at that given moment? Consider any particular driver assigned to the slow route in the earlier example—he could always "defect" to the fast route and reduce his driving time, so why wouldn't he? And if everyone did this, they all revert to the competitive equilibrium of existing apps.

If we think about it for a moment, it seems possible that even current navigation apps such as Google Maps and Waze could also be susceptible to various kinds of cheating or manipulation. For example, I could lie to Waze about my intended origin and destination, in an effort to influence the routes it recommends to other users in a way that favors me—creating false traffic that causes the Waze solution to route other drivers away from my true intended route. Manipulation of this variety apparently occurred in residential Los Angeles neighborhoods frustrated by the amount of Waze-generated traffic, as reported by the *Wall Street Journal* in 2015:

> Some people try to beat Waze at its own game by sending
> misinformation about traffic jams and accidents so it will steer
> commuters elsewhere. Others log in and leave their devices in

their cars, hoping Waze will interpret that as a traffic standstill and suggest alternate routes.

But the incentive problems that Maxwell faces are arguably even worse, because they are not simply about drivers lying to the app; rather, the problem is drivers disregarding its recommendations entirely when they are not best responses.

There are a couple of reasonable replies to this concern. The first is that we may eventually (perhaps even soon) arrive at an era of mostly or even entirely self-driving cars, in which case the Maxwell solution could simply be implemented by centralized fiat. Public transportation systems are generally already designed and coordinated for collective, not individual, optimality. If you want to fly commercially from Ithaca, New York, to the island town of Lipari in Italy, you can't simply direct American Airlines to take a nonstop route along the great circle between the two locations—instead you'll have multiple flight legs and layovers, all for the sake of macroscopic efficiency at the expense of your own time and convenience. In a similar vein, it would be natural for a massive network of self-driving cars to be coordinated so as to implement navigation schemes that optimize for collective average driving time (and perhaps other considerations, such as fuel efficiency) rather than individual self-interest.

But even before the self-driving cars arrive en masse, we can imagine other ways Maxwell might be effectively deployed. One is that if, as in our two-route example above, Maxwell randomly chooses the drivers who are given nonselfish routes, users might have a stronger incentives to use the app, since over time the assignment of nonselfish routes will balance out across users, and then each individual user would enjoy lower average driving time. So while you might have an incentive to disregard Maxwell's recommendation of a slower route on any given trip (which you might well discover by using Google Maps to see your selfish best-response route), you know that over time you will benefit from following Maxwell's suggestions (as long as others do as well). We might call this phenomenon cooperation through averaging, which is also known to occur when human subjects play repeated rounds of Prisoner's Dilemma. But perhaps there is a better and more general solution to these incentive concerns.

COOPERATION THROUGH CORRELATION

Let's review where we are. Maxwell may have a better collective solution, but it is vulnerable to defection, and even the selfish navigation apps may be prone to manipulation. Both approaches have good algorithms, but the concern is that their goals can be compromised by human nature.

It turns out that sometimes these concerns can be overcome by considering yet a third notion of solution in games (our first being the selfish equilibrium, and the second being the best social welfare but non-equilibrium Maxwell solution). This third notion is known as a correlated equilibrium, and it too can be illustrated by a simple situation involving driving. Imagine an intersection of two very busy roads, one of which has a yield sign and the other of which does not. Then not only the law but also the selfish equilibrium is for drivers on the yielding road to always wait for an opening before continuing, and for drivers on the through road to speed along. Given what the drivers on the other road are doing, everyone is following their best response. But drivers on the yielding road suffer all the waiting time, which might feel unfair to them.

In this example a correlated equilibrium could be implemented with a traffic signal, which now allows drivers to follow strategies that depend on the signal, such as "If the light shows green to me, I will speed through, and if it shows red to me, I will wait." If everyone follows this strategy, they are all best-responding, but now the waiting time is split between the two roads—a fairer outcome not possible with only yield signs. The traffic signal is thus a coordination (or correlating) device that allows cooperation to become an equilibrium.

Can cooperation via coordination help solve Maxwell's incentive problems? The answer is yes—at least in principle. Very recent research has shown how it is possible to design a variant of Maxwell's algorithm—let's call it Maxwell 2.0—that quickly computes a correlated equilibrium, and enjoys three rather strong and appealing incentive properties. First, it is in the best interest of any driver to actually use Maxwell 2.0: nobody has an incentive to opt out and use another app instead (unlike Maxwell 1.0). Second, it is in the best interest of any driver to honestly input his true origin and destination: one cannot beneficially manipulate the solution found by Maxwell 2.0 by lying to it. (This is a property known as "truthfulness" in game theory.) Third and finally, it is in the best interest of any driver to actually follow the route recommended by Maxwell 2.0 after he sees it. So all drivers want to use Maxwell 2.0, and to use it honestly—both in what they input to it and in following its output.

How does Maxwell 2.0 achieve these apparently magical properties? Looking back to Chapter 1, it does so by applying differential privacy to the computation of the recommended routes in a correlated equilibrium. Recall that differential privacy promises that the data of any individual user cannot influence the resulting computation by very much. In this case a user's data consists of both the origin and destination he reports to Maxwell 2.0 and the traffic data his GPS locations contribute. The computation in question is the assignment of driving routes in a correlated equilibrium. Since a single driver's data has little influence, it means manipulations like lying about where you want to go or leaving the app on in your parked car won't benefit you or change what others do. And since a correlated equilibrium is being computed, your best response is to follow the suggested route.

Note that there was no explicit privacy goal here. Rather, the incentive properties we desired were a by-product of privacy. But at a high level, it makes sense: if others can't learn anything about what you have entered into Maxwell 2.0 or what it told you to do, then you can't beneficially manipulate your inputs to change the behavior of others. That techniques developed for one purpose (like privacy) turn out to have applications elsewhere (like incentivizing truthfulness) is a common theme in algorithms. In fact, differential privacy has many other non-privacy-related applications—we will see another one in Chapter 4.

GAMES EVERYWHERE

Even though Maxwell is only a hypothetical app (at least for now), we spent some time on the commuting game because it crisply illustrates a number of more general themes, and does so in a situation with which many of us have daily experience. These themes include:

- Individual preferences (e.g., where you want to drive from and to) that may be in conflict with those of others (e.g., traffic).
- The notion of a competitive or "selfish" equilibrium, and the observation that convenient modern technology (e.g., Waze) might drive us toward this equilibrium.
- The observation that there might be socially better outcomes that can be found with fast algorithms (e.g., Maxwell) that also enjoy good incentive properties (e.g., Maxwell 2.0).
- The lesson that when an app is mediating or coordinating the preferences of its users (as opposed to simply using their data for some other purpose, such as building a predictive model), the algorithm design must specifically take into consideration how users might react to its recommendations—including trying to manipulate, defect, or cheat.

In the rest of this chapter, we'll see that these same ideas apply in a wide variety of other modern, technology-mediated interactions, from routine activities such as shopping and reading news to more specialized situations such as assigning graduating medical students to hospital residencies, and even to kidney transplants. In some cases we'll see that the algorithms involved might be pushing us toward a bad equilibrium, and in other cases we'll see they are doing social good. But in all of them, the design of the algorithm and the preferences and desires of the users are inextricably intertwined.

SHOPPING WITH 300 MILLION FRIENDS

Like driving, shopping is another activity that many of us engage in daily and that has been made more social and game-theoretic by technology. Before the consumer Internet, shopping—whether for groceries, plane tickets, or a new car—was largely a solitary activity. You went to physical, local stores, and your decisions were based on your own experience and research (and perhaps the advertising you were exposed to). For bigger purchases such as a car or a television, there might be publications such as *Consumer Reports*. But for most things, you were more or less on your own. As in the commuting game, you had shopping preferences—admittedly more complex, multifaceted, and harder to articulate than simply wanting to drive from point A to point B. But there were very few tools to help you optimize your decisions. It was the shopping equivalent of the era of spotty traffic reports and fold-up maps.

As readers will have experienced, all of this changed with the explosive growth of online shopping. Once we began researching and purchasing virtually everything imaginable on the web, we provided retailers such as Amazon extremely fine-grained data on our interests, tastes, and desires. And as we have discussed in previous chapters, machine learning could then take this data and build detailed predictive models that generalized from the products and services we already did like to the ones we would like if only we were made aware of them. The technical term in the computer science community for this general technology is *collaborative filtering* (which was widely used in the Netflix competition, whose privacy violations were discussed in Chapter 1). It's worth under-standing a bit about the algorithms and models involved, and how they have become more sophisticated and powerful over time.

First of all, the "collaborative" in collaborative filtering refers to the fact that not only will your own data be used to make recommendations to you, but so will everyone else's data. The most rudimentary techniques rely just on counting, and appear in the form of Amazon's "customers who bought this item also bought" suggestions. By simply keeping statistics on items frequently purchased together or in quick succession by other users, Amazon can tell you what you might want or need with your current pur-chase. These kinds of counting recommendations aren't really generaliz-ing in any deep or meaningful way—one just needs to look up the histor-ical frequency statistics—and they tend to be relatively obvious, like tennis balls to go with your new racquet, or a David Foster Wallace novel to go with your purchase of Thomas Pynchon's *Gravity's Rainbow*.

What other items do customers buy after viewing this item?

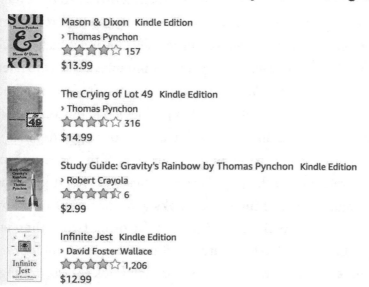

Mason & Dixon Kindle Edition
› Thomas Pynchon
★★★★☆ 157
$13.99

The Crying of Lot 49 Kindle Edition
› Thomas Pynchon
★★★☆☆ 316
$14.99

Study Guide: Gravity's Rainbow by Thomas Pynchon Kindle Edition
› Robert Crayola
★★★★☆ 6
$2.99

Infinite Jest Kindle Edition
› David Foster Wallace
★★★★☆ 1,206
$12.99

Fig. 19. Amazon recommendation for products related to the Thomas Pynchon novel *Gravity's Rainbow,* based on simple customer purchase statistics.

This level of collaborative filtering might capture similar or related products but has no explicit notion of similar *people*. Instead of just suggesting balls to go with racquets, how can we solve harder problems, like suggesting vacation destinations to someone who has only ever purchased books online? If your taste in reading reveals enough about what type of person you are, then we can suggest the vacations taken by others like you.

SHOPPING, VISUALIZED

What's a principled, algorithmic way of mapping users to their "types" based on their shopping history, and how would we discover these types in the first place? Let's imagine a greatly simplified world in which there are only three products for sale on Amazon, all of them books: Thomas Pynchon's *Gravity's Rainbow*, David Foster Wallace's *Infinite Jest*, and Stephen King's *The Shining*. Suppose there are one thousand Amazon users, each of whom has rated all three of these books on a continuous scale of 1 to 5 stars. (We'll use a continuous scale—which allows ratings like 3.729—because it will make visualization easier, but everything we'll say applies equally well to the discrete rating systems common on platforms like Amazon.) Then we can actually plot and visualize each user as a point in three-dimensional space. For instance, a user who rated *Gravity's Rainbow* as a 1.5, *Infinite Jest* as a 2.1, and *The Shining* as a 5 would have an x coordinate value of 1.5, a y value of 2.1, and a z value of 5. Together the one thousand users give us a cloud of points in space.

What might we expect this cloud to look like? If there were absolutely no correlations in the population ratings—that is, if knowing a user's ratings for one of the three books gives us no information or insight as to how much they might like the other two—then we would expect this cloud to look rather diffuse and spread out. For instance, if there are no correlations, and for each book the user ratings are evenly distributed between 1 and 5, the cloud would essentially fill up the three-dimensional space, as in the top panel of Figure 20.

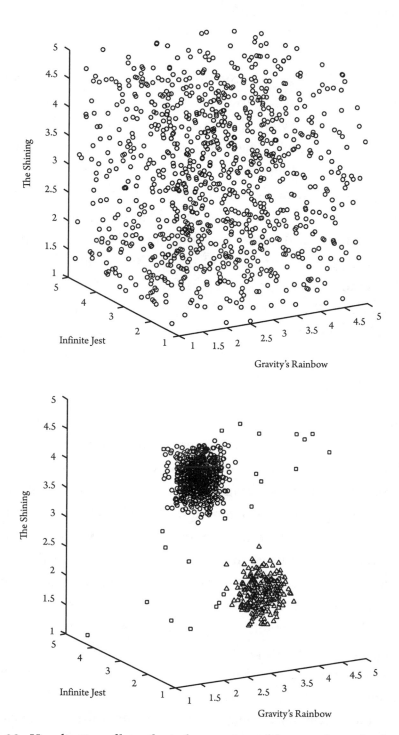

Fig. 20. Visualization of hypothetical user ratings of three products. On the left, ratings across products are uncorrelated, and each product is equally likely to be rated anywhere from 1 to 5. On the right, there are strong correlations across products, and some products have higher average ratings than others.

But we might well expect the cloud to look quite a bit less random and more structured. In particular, suppose it's the case that readers of *Gravity's Rainbow* are, on average, much more likely to enjoy *Infinite Jest* than *The Shining*; conversely, fans of *The Shining* tend to find both *Gravity's Rainbow* and *Infinite Jest* a bit obtuse and over-wrought. Then the ratings would actually form two distinct clouds or clusters of points—one cloud with high ratings for *The Shining* and low ratings for the other two (the cloud of circles in the bottom panel of Figure 20), and the other with high ratings for *Gravity's Rainbow* and *Infinite Jest*, and low ratings for *The Shining* (the cloud of triangles in the bottom panel of Figure 20). The structure in this example jumps right out at us even though there are some users (the squares) that don't really fall into either of the two clouds; they are outliers that are not as easily categorized.

Perhaps unsurprisingly, the real world looks quite a bit more like the bottom panel than the top panel—as we have mentioned elsewhere, there really are very strong correlations in the products people buy or rate (as well correlations between more abstract things, such as between our political beliefs, race, religious practices, and sexual preferences). Before turning to the question of how we might identify these clouds algorithmically, let's appreciate two very powerful properties they enjoy.

First, suppose we are given a user who has only rated *Infinite Jest* and given it a 4.5. We can't map this user into our three-dimensional space because we are missing two of the ratings. But if we instead want to *recommend* a new book to this user, the data makes it intuitively clear what we should do: when we project the data onto the *Infinite Jest* axis, it is quite likely that this user is in reality a member of the triangle cloud and not the circle cloud and thus would enjoy *Gravity's Rainbow* much more than *The Shining*. So knowing about the two clouds lets us generalize, or infer what new products this user might like. (If instead we just know a user liked *The Shining*, we don't have a book to recommend, but at least we know two books *not* to recommend.)

The second nice property is that the data itself automatically tells us what the user "types" are. We don't have to guess or posit the existence of prefabricated categories such as "postmodern fiction lover" or "horror fan," or identify books as postmodern or horror. We don't even have to give names to the types at all—they are just groups of users whose tastes across products tend to be similar.

Incidentally, the attentive reader might note than in our three-book example, it's not obvious why it's helpful to identify user types or clusters, instead of just making the simplistic product correlation recommendations discussed earlier ("People who bought *Gravity's Rainbow* often bought *Infinite Jest*"). That's because the benefits of user modeling really arise when there are far more products than just three. Say you've rated or bought one hundred products on Amazon so far. Instead of picking one of your purchases and suggesting a related product, we use all one hundred of your purchases to identify your type first, and then use your type to suggest things you might like from the entire universe of products. This is how it is possible to recommend vacation destinations, rather than just books, to someone who has only ever purchased books.

A DIFFERENT KIND OF CLOUD COMPUTING

In the example above, simple visualization of the data made the two clouds or types of users jump out at us—we could just "eyeball" them. But not only will this not scale up when we have 500 million products (the approximate number sold on Amazon) instead of three, it's not even well-defined. We need a more algorithmic specification of the problem.

Let's still represent users as points in product rating space, but now that space will be much higher-dimensional—perhaps not all 500 million Amazon products, but maybe a very large sample of diverse and representative ones. Given our cloud of user rating points, one natural formulation of our problem is the following:

Find a division of users into K groups such that the average distance between users in the same group is much smaller than the average distance between users in different groups.

In our simple example above, if we choose $K = 2$, it's pretty clear that the groups that optimize this criterion are the circle and triangle groups we identified; for completeness we need to assign the outlier black points to whichever cloud they are closest to, circle or triangle. But if outliers are rare, they won't influence the optimal solution very much. Furthermore, at least in this extreme example, the data also tells us what the "right" value of K is—if we increase to $K = 3$, we'll perhaps either put a square outlier in its own private cloud, or maybe split the circle or triangle cloud in half, but neither of these would improve our solution much, so we should stop at $K = 2$.

For a large number of users in a large space of products, the problem we have identified above can be computationally challenging to solve optimally or exactly, but there are extremely efficient heuristics that find good approximate solutions. A naive approach would be to start by picking K random center points in the product rating space and assign every user to the nearest of these K points. This will give us an initial division of users into groups that will probably do terribly on the criterion we've specified (make within-group distances small compared to between-group distances). But now we can gradually make small adjustments to these K centers to improve the criterion until we get stuck and can't make any further improvement; in the language of algorithms, this would produce a locally optimal solution, but possibly not a globally optimal one. Once we have produced the groups or clusters, we can always produce a canonical user or type for each group— for instance, by averaging the ratings of the users in that group.

There are many machine learning algorithms that work much faster and better than this naive approach; they have rather technical names such as K-means, expectation-maximization, and low-rank matrix approximation. But for our purposes, they all share the same top-level goals: given a colossal and incomplete dataset of user product ratings or purchases (incomplete because virtually every user has purchased only a small fraction of Amazon products), discover a small number of canonical user types, which can then be used to make accurate recommendations of new products to users based on their type.

THE ECHO CHAMBER EQUILIBRIUM

So just like Waze, Amazon gathers all of our collective behavioral data (about shopping, instead of driving and traffic) and uses it to try to optimize recommendations for each of us individually (about what products to buy, instead of what route to drive). What's initially harder to see in this analogy is the sense in which shopping is a "game" the way commuting is, and whether Amazon is nudging its users toward a bad selfish equilibrium. Everyone else driving on the roadways clearly impacts you negatively in the form of traffic. But what consequences, good or bad, does everyone else's online shopping have for you?

From our description of collaborative filtering, it's at least clear there *are* consequences. When our collective data is used to estimate a small number of user types and you are assigned to one of them, the products suggested to you are narrowed by the scope of the model created, which is a function of everyone else's shopping activity. If the model implicitly learns that people who drive American cars and enjoy hunting are more likely to read Stephen King than postmodern fiction, they will be steered in that direction, even though they might have enjoyed Thomas Pynchon. Of course, people still have free will and can choose to ignore Amazon recommendations in favor of their own research and wanderings. But this is true of Google Maps and Waze as well— one can always choose the slow or scenic route instead of the time-saving selfish one, or opt out of using such apps at all. But the more of us that adopt their suggestions, the more our collective behavior is influenced or even determined by them.

Perhaps in the case of shopping, we can argue we are being nudged toward a good or at least neutral equilibrium—one in which we all benefit from the insights gleaned by machine learning from our aggregate purchases, in the form of individually optimized recommendations for new products. Since we're not all competing for a limited resource such as the capacity of the roadways, maybe all of us can simultaneously optimize without making anyone else worse off, which is unavoidable in commuting.

But we might feel differently about the same methods applied to a very similar problem, which is news filtering. Platforms such as Facebook also apply the same powerful machine learning techniques to build individual profiles of user interests based on collective data, and use these models to choose what appears in their News Feed. The aforementioned narrowing that can occur in this process has oftentimes been referred to as an "echo chamber" (or "filter bubble")—meaning that users are being shown articles and information that align with or echo their existing beliefs and prejudices. And what articles appear on a user's News Feed can further reinforce the existing behavior that led the algorithm to select them in the first place, resulting in another feedback loop.

In game-theoretic terminology, these services may have led us to a bad equilibrium in which politics and public discourse have become polarized, and which encourages all of us to become less informed about, and less tolerant of, opposing perspectives. We're all best-responding—at any given moment, we'd prefer to read an article confirming rather than challenging our beliefs—but perhaps as a society we have been led to a less healthy place by technology. This polarization has exacerbated the effectiveness of "fake news" inside online communities that are rarely exposed to dissenting information. There have also been widespread incidents of deliberate manipulations, such as bogus Facebook accounts—complete with user profiles seeded with specific identities, activities, and opinions—designed so that the fake users will have their posts show up in the News Feeds of particular online communities or echo chambers.

Note that the Amazon and Facebook equilibria are largely the same, and both are the direct result of algorithmic, model-based systems trying to simultaneously optimize everyone's choices. It's just that when an informed and deliberative society is at risk, the selfish equilibrium feels much worse than when we're all just choosing our next novel or vacation. As with fairness (Chapter 2), the stakes matter.

QUANTIFYING AND INJECTING DIVERSITY

If we don't like the echo chamber equilibrium for news filtering, or maybe even for shopping, how might we improve the algorithms used by platforms like Facebook and Amazon? One natural approach is to increase the diversity or exploration in the recommendations made to users—to deliberately expose people to news articles or products that differ from their obvious preferences and past behavior. Furthermore, this need not be done in a random or haphazard fashion (which might strike users as simply irrelevant and irritating); it can be done algorithmically.

Consider the mapping of users to types implemented by methods such as collaborative filtering. Because such models position users and types in space, there is also a quantitative measure of distance between them. So not only can the model estimate your type, it also knows how similar or different you are from all of the other types. For example, in our three-book example above, we can measure the distance between the centers of the circle cloud that likes *The Shining* and the postmodern fiction triangle cloud.

If we wanted news filtering algorithms that really challenged your worldview, we could deliberately pepper you with articles most aligned with your opposite type—the one farthest in space from your own. Of course, this might be too much too soon, and simply alienate or even offend users. But the point is that such algorithms can provide a "knob" (akin to the fairness and privacy parameters discussed in earlier chapters) that can be adjusted (even by individual users, if platforms so chose) from purely selfish echo chamber behavior, to making recommendations a bit outside your comfort zone, to really showing you how your polar opposites see the world. And of course if want to be algorithmically transparent, we can always choose to put the exploratory recommendations in a clearly marked "Opposing Viewpoints" sidebar.

So unlike our proposals for the commuting game—which required us to adopt different and more speculative algorithms (like Maxwell 2.0) rather than the selfish one in order to achieve socially better solutions—we can attempt to address the echo chamber equilibrium with algorithms and models that are only minor variants of the ones currently deployed. It might just involve changing a few lines of existing code.

MEDICAL MATCHMAKING

Throughout this chapter, we have deliberately chosen to illustrate the interplay between individual preferences, collective welfare, game theory, and algorithms through examples with which most of us have daily experience, such as driving, shopping, or reading news. But there are many more specialized settings in which algorithmic game theory has long played a central role in highly consequential decisions.

One such setting is broadly known as *matching markets* in economics. While this phrase may bring to mind dating apps like Coffee Meets Bagel, matching markets are usually found in much more formal scenarios in which we want to pair up individuals with each other, or individuals with institutions. One long-standing application area is medical residency hiring, in which the approach we'll describe is implemented as the National Resident Matching Program (NRMP, affectionately known as "The Match").

The basic problem formulation is as follows. Candidates for medical residencies each have an individual ranking of potential programs. For example, suppose the candidates Elaine and Saeed have the following ranked list of residencies (ignore the annotating characters for now, which we discuss shortly):

Elaine	Saeed
Harvard	Cornell
Johns Hopkins	UC San Diego @
UC San Diego &	Harvard #
Baylor	Johns Hopkins

Based on application materials and interviews, the schools of course also have their own ranked lists of candidates, such as:

Harvard	UC San Diego
Saeed #	Roger
Elaine	Saeed @
Roger	Elaine &
Gwyneth	Mary

So this is a two-sided market—candidates and hospitals—and there are also capacity constraints, because each candidate can of course take only one residency, and each hospital can only take a limited number of residents (which for simplicity we'll assume is also just one). Thus, as with dating and commuting, we once again have a large system (many thousands of applicants, and hundreds of schools) of interacting, competing preferences and would like to specify a desirable solution—and a fast algorithm for finding one.

Let's approach the notion of a desirable solution by first specifying what we *don't* want to happen. Consider candidates Elaine and Saeed in the example above. Suppose we match Elaine with UC San Diego (indicated by the "&" characters next to both), and Saeed with Harvard (indicated by the "#" characters next to both). Then regardless of any other matches we make, this solution is *unstable*, because Saeed prefers UC San Diego to Harvard, and UC San Diego prefers Saeed to Elaine—that is, the match indicated by the "@" characters would be preferable to both Saeed and UC San Diego, compared to their respective outcomes under the & and # matches. So while Elaine and Harvard might be happy, Saeed and UC San Diego have an incentive to deviate or defect from their assigned matches (Harvard and Elaine, respectively) and hook up with each other instead. A solution in which there are no such potential defections is called a stable matching. A matching with this property is not at risk of unraveling as students and hospitals iteratively defect from their proposed matches.

A stable matching is conceptually quite similar to a Nash equilibrium, but now two parties (a candidate and a medical school) must jointly defect to a mutually preferred outcome, due to the two-sided nature of the market. And like a Nash equilibrium, a stable matching in no way promises that everyone will be satisfied with the outcome: a candidate assigned to her 117th-favorite hospital may not be happy, but as in a Nash equilibrium, there is nothing she can do about it, because the 116 hospitals she prefers already have candidates they like better than her. Candidates and hospitals that are paired together in a stable matching are stuck with each other. Nevertheless, a stable matching is an intuitive solution to such pairing or assignment problems—certainly any solution that is *not* a stable matching is vulnerable to defections and is therefore problematic—and it is a *Pareto optimal* solution as well, in the sense that there is no way to make anyone better off without making someone else worse off (similar to the accuracy-fairness Pareto curves discussed in Chapter 2).

Like other topics in this chapter, stable matchings have both a long history and fast algorithms for computing them, going back at least to the seminal 1962 work of David Gale and Lloyd Shapley. The so-called Gale-Shapley algorithm is sufficiently simple that it even has a plain English description on Wikipedia, whimsically phrased as pairing men and women via traditional Victorian courtship:

- In the first round, *a*) each unengaged man proposes to the woman he prefers most, and then *b*) each woman replies "maybe" to her suitor that she most prefers and "no" to all other suitors. She is then provisionally "engaged" to the suitor she most prefers so far, and that suitor is likewise provisionally engaged to her.
- In each subsequent round, *a*) each unengaged man proposes to the most-preferred woman to whom he has not yet proposed (regardless of whether the woman is already engaged), and then *b*) each woman replies "maybe" if she is currently not engaged or if she prefers this suitor over her current provisional partner (in this case, she rejects her current provisional partner who becomes unengaged). The provisional nature of engagements preserves the right of an already-engaged woman to "trade up" (and, in the process, to "jilt" her until-then partner).
- This process is repeated until everyone is engaged.

The Gale-Shapley algorithm has two very nice properties. First, regardless of the preferences, everyone gets matched (if there are equal number of men and women, and they don't deem any of their potential partners as absolutely unacceptable; for instance, every medical student wants to do a residency no matter what). Second, the matching computed by the algorithm is stable in the sense we described above. Algorithms generalizing to the cases where there are unequal numbers of men and women (or students and medical schools), or where one side of the market can accept more than one partner (as in medical residencies), also exist. These algorithms are in widespread practical use, including in assigning actual medical residencies and other competitive admissions settings, such as matching students to public high schools and matching pledges to sororities in universities. (In contrast, US undergraduate college admissions are generally done in a much more haphazard fashion, giving rise to experiments in admissions office gamesmanship such as early decision, early action, requiring standardized tests or not, additional essays, and the like—all to the exhaustion and frustration of applicants and their parents.)

One of the most striking applications of algorithmic matching in the real world—one that literally has saved human lives—is to the problem of paired kidney donation. Many people with kidney disease die every year while awaiting a transplant donor, and the problem is exacerbated by the fact that the blood type of a donor must be compatible with that of the recipient to be biologically viable (there are also a variety of other medical compatibility constraints). We can view the blood type and biology of a donor as a form of "preferences" over recipients—a donor "prefers" to donate to compatible recipients, and not to incompatible ones. Similarly, a recipient prefers to receive a transplant from a compatible donor.

While there are many details that make this problem more complicated than medical residency matching, there are again practical, scalable algorithms that maximize the efficiency of the solution found, where here efficiency means maximizing the total number of compatible transplants that occur globally—ideally across all hospitals, not just within a single one. For his algorithmic and game-theoretic insights on this problem (and the others we have mentioned, including the medical residency match) and his efforts to convince the medical community and hospitals that it was worth the effort to pool their transplant donors, recipients and data, Alvin Roth was awarded the 2012 Nobel Prize in economics—along with the aforementioned Lloyd Shapley, whose early work initiated the era of algorithmic matching.

ALGORITHMIC MIND GAMES

We've now seen a variety of modern settings (and potentially future ones, such as self-driving cars) in which game-theoretic modeling can provide both conceptual guidance and algorithmic prescriptions to problems in which a large number of individuals or institutions have complex and potentially competing preferences. The proposed algorithms are socially aware in the sense that they attempt, albeit imperfectly, to mediate these preferences in a way that has socially desirable properties such as efficiency (e.g., low collective commute time), diversity (e.g., of news exposure), or stability (e.g., of matchings).

A rather different and more recent use of game theory is for the internal design of algorithms, rather than for managing the preferences of an external population of users. In these settings, there is no actual human game, such as commuting, that the algorithm is helping to solve. Rather, a game is played in the "mind" of the algorithm, for its own purposes.

An early example of this idea is self-play in machine learning for board games. Consider the problem of designing the best backgammon-playing computer program that you can. One approach would be to think very hard about backgammon strategy, probabilities, and the like, and hand-code rules dictating which move to make in any given board configuration. You would try to articulate and impart all of your backgammon wisdom in computer code.

A rather different approach would be to start with a program that knows the rules of backgammon (i.e., which moves are legal in a given configuration) but initially knows absolutely nothing about backgammon strategy (i.e. how to play well, rather than just legally). The initial version of this program might simply choose randomly among its legal moves at every round—surely a losing strategy against even a novice player. But if we make this initially naive program adaptive—that is, if its strategy is actually a model mapping board configurations

to next moves that can be tuned and improved with experience—
then we can take two copies of this program and have them improve
by playing *each other*. In this way we turn playing backgammon into a
machine learning problem, with the requisite data being provided by
simulated self-play.

This simple but brilliant idea was first successfully applied by Gerry
Tesauro of IBM Research in 1992, whose self-trained TD-Gammon
program achieved a level of play comparable to the best humans in
the world. (The "TD" stands for "temporal difference," a technical term
referring to the complication that in games such as backgammon,
feedback is delayed: you do not receive feedback about whether each
individual move was good or bad, only whether you won the entire
game or not.) In the intervening decades, simulated self-play has proven
to be a powerful algorithmic technique for designing champion-level
programs for a variety of games, including quite recently for Atari video
games, and for the notoriously difficult and ancient game of Go.

It might seem natural that internal self-play is an effective design principle if your goal is to actually create a good game-playing program.
A more surprising recent development is the use of self-play in algorithms whose outward goal has nothing to do with games at all. Consider
the challenge of designing a computer program that can generate realistic but synthetic images of cats. (We'll return shortly to the question of
why one might be interested in this goal, short of simply being a cat
fanatic.) As with backgammon, one approach would be the knowledge-
intensive one—we'd gather cat experts and study images of cats in order
to understand their coloring, physiology, and poses, and try to somehow
encode all of this expertise in a program that generated random, photo-
realistic cat images. It seems difficult to even know where to begin.

Or we could again take the simulated self-play approach. The
high-level idea is to invent a game between two players that we'll call
Generator and Discriminator. The goal of Generator is to create or
generate good artificial cat images. Discriminator is given a sample

of fake cats created by Generator, as well as a large collection of real cat images, and has the goal of reliably discriminating between the fake and real cats. Discriminator can actually be a standard machine learning algorithm whose training data labels the real cats as positive examples and the fake ones as negative examples. But as with TD-Gammon, it is crucial that *both* players are adaptive.

At the start of the game, Generator (who does not even have the luxury of seeing any real cat images) plays terribly, creating pictures that look like random collections of pixels. Thus Discriminator's task is comically easy—just differentiating between real cats and garbled bits. But after the first round, once Discriminator has committed to its first model, Generator uses this model to modify its fake cats in a way that makes Discriminator slightly more confused about the real versus the fake . . . which in turn forces Discriminator to revise its model to solve this slightly harder problem. We continue this back-and-forth ad infinitum. If both players become experts at their respective tasks, Generator creates incredibly realistic cats, which the Discriminator might still be able to distinguish from the real thing better than humans could. But if the Generator is sufficiently good at learning, this is clearly a losing game for the Discriminator.

The technical name for the algorithmic framework we have been describing is a generative adversarial network (GAN), and the approach we've outlined above indeed seems to be highly effective: GANs are an important component of the collection of techniques known as deep learning, which has resulted in qualitative improvements in machine learning for image classification, speech recognition, automatic natural language translation, and many other fundamental problems. (The Turing Award, widely considered the Nobel Prize of computer science, was recently awarded to Yoshua Bengio, Geoffrey Hinton, and Yann LeCun for their pioneering contributions to deep learning.)

Fig. 21. Synthetic cat images created by a generative adversarial network (GAN), from https://ajolicoeur.wordpress.com/cats.

But with all of this discussion of simulated self-play and fake cats, it might seem like we have strayed far from the core topic of this book, which is the interaction between societal norms and values and algorithmic decision-making. However, recent research has shown that these same techniques can in fact play a central role in the design of better-behaved algorithms as well.

For example, recall our discussion of "fairness gerrymandering" back in Chapter 2—the phenomenon that despite building a model that does not discriminate by gender, race, age, disability or sexual orientation separately, we nevertheless end up with one that is unfair to gay disabled Hispanic women over age fifty-five making less than $50,000 a year. This was just another instance of machine learning not giving you "for free" something you didn't say you wanted. We mentioned briefly in that chapter that one solution to this problem involved an algorithm that simulated play between a Learner who would like to minimize error and a Regulator who continually confronts the Learner with subgroups suffering discrimination under the Learner's current model. This is another example of simulated game play as design principle, with the Regulator taking the place of a backgammon program or a Generator. Here the Regulator's goal (fairness) may be in conflict with the Learner's (accuracy), and the outcome (which is actually a Nash equilibrium of a precisely defined game) will be a compromise of the two—as desired.

Similarly, game-theoretic algorithm design has also proven useful in differential privacy. For example, while there may not be much motivation to generate fake cat pictures—we have plenty of the real thing—there is very good reason to generate realistic-looking but fake or synthetic medical records. In this case, the real thing can't generally be widely shared because of privacy concerns—often to the detriment of scientific research. One recent application of GANs is to the generation of highly realistic sets of medical records that can be made publicly available for research purposes while preserving the individual privacy of the patients whose real records trained the GAN. This is again achieved by framing the algorithm as a game—here between a Generator who desires to keep the synthetic dataset as close to the real dataset as possible and a Discriminator who wants to point out differences. So long as the Discriminator is differentially private, the synthetic data produced by the Generator will be as well. So while GANs are in their relative infancy, scientists are starting to find compelling (non-feline) uses for them.

GAMES SCIENTISTS PLAY (WITH DATA)

Most of this chapter examined settings in which people have preferences that may conflict with each other, and the ways in which algorithms can play a role in their mediation and management (for better or worse). Many of these "games" involved everyday activities such as driving or shopping, and some were more rarified, like assigning medical residencies.

We have one last case study in which the participants may not think of themselves as players in a complex game—but they are. It's the realm of modern scientific research, especially in the rapidly expanding disciplines in which data analysis and predictive modeling are playing an increasingly dominant role (which of course includes machine learning itself). The players in this game are the professors, graduate students, and industry researchers in data-driven fields. Their incentives involve the publication of novel and influential results, often requiring improvement in quantitative metrics such as the error rate on benchmark data sets or in the findings of prior experiments and analyses. Each new publication influences subsequent data-gathering, modeling, and choices made by the scientific community. It's a game where even individually careful scientists can participate in a bad equilibrium in which there is collective overfitting to widely used data sets—thus leading to spurious and false "scientific" findings.

It's a game that deserves a chapter of its own.

4

Lost in the Garden

Led Astray by Data

PAST PERFORMANCE IS NO GUARANTEE OF FUTURE RETURNS

Imagine that you wake up one day and check your email. Waiting for you in your inbox is a message with the subject line "Hot Stock Tip!" Inside, you find a prediction: shares of Lyft, the recently listed ride-sharing company (NASDAQ:LYFT), are going to end the day up. You should buy some now! Of course, you don't take this advice— how did it get past your spam filter? But the prediction is specific enough that you remember it, and you check Google Finance after the close of the market. Sure enough, LYFT shares ended the day up. Amusing, but not terribly surprising—if the sender had merely flipped a coin and guessed, he would have been right about the direction of the stock half the time.

The next day, you get another email from the same person. It tells you that today LYFT is going to end the day down, and you should short-sell it. Of course, you don't—but you take note. And at the end of the day, you are amused to find that again the sender was right, and LYFT was down more than 5 percent. The next day you get another email saying LYFT will go down again. The next day, yet another— LYFT will end the day up. This goes on for ten days, and every day the sender correctly predicts the direction of the stock.

At first you were just curious, but now you are starting to really pay attention. After a few days you suspected that the sender might be a Lyft employee who is giving you illegal tips based on inside information. But you then discovered that there's really not been any news or events to support this hypothesis, or to explain the seemingly random movements of LYFT. Finally, on the eleventh day, the sender emails you a request. He wants you to pay him to continue giving you stock recommendations. He will charge a high fee, but it will be worth it, he says. After all, he's already given you a free demonstration of how talented he is at predicting the movement of stocks.

You take a moment to consider it. Should you view his demonstrated streak of ten correct predictions as convincing? Since you have some scientific training, you decide to formulate a null hypothesis and see if you can convincingly reject it. Your null hypothesis is that this fellow is no better at predicting stock movements than the flip of a coin—on any particular day, the probability that he correctly guesses the directional movement of LYFT is 50 percent. You go on to compute the p-value corresponding to your null hypothesis—the probability that if the null hypothesis were true, you would have observed something as extreme as you did: ten correct predictions in a row. Well, if the sender had only a 50 percent chance of getting the answer right on any given day, then the chance that he would get it right ten days in a row—the p-value—would be only about .0009, the probability of flipping a coin ten times in a row and getting heads each time. This is very small—well below the .05 threshold for *p*-values that is often taken as the standard for statistical significance in the scientific literature. So, marshaling all of your scientific training, you decide to forget your skepticism and reject the null hypothesis: you decide your interlocutor is actually pretty good at predicting stock movements. You send him the money he has asked for. As promised, you keep getting daily stock tips from him, but now something seems different: suddenly his predictions are as likely to be wrong as they are to be right. Now he's no better than a coin flip.

What went wrong? How did your scientific training fail here? The keys to this email scam are *scale* and *adaptivity*. What you failed to take into account is the fact that the sender has sent emails not just to you but to many other people as well. Here is how it works. On the first day, the scammer sends out emails to 1 million people; this is the scale part. Half of the emails predict that the stock will go up. Half predict that the stock will go down. No matter what happens, the predictions sent to half of the recipients will turn out to be correct. The scammer never sends another email to the people to whom he sent incorrect predictions; he's lost their faith already. The next day, he sends emails to the 500,000 people to whom he sent a correct prediction on the first day. Again, to half of these he predicts that the stock will go up, and to half he predicts it will go down. He continues in this way, each day sending a correct prediction to half of his targets and discarding those recipients to whom he made an incorrect prediction; this is the adaptivity part.

At the end of ten days, there are about 1,000 people left who have received ten correct predictions in a row. To these people—and only to these people—does he send the final request for money. Each of these people might individually reason that the chance that the predictions could have been made correctly so many times in a row is tiny—but the truth is that no matter what happened in the stock market, the scammer was *guaranteed* to be able to demonstrate this kind of miraculous performance to around 1,000 people. And given that you received the final request for money, the chance that the scammer got all ten "predictions" correct wasn't tiny—it was certain.

It is helpful to visualize this scam as an inverted "tree" of possible outcomes. At the top or root of this tree are the initial 1 million potential victims of the scam. These are randomly divided into the 500,000 that receive an "up" prediction on the first day (represented as the left branch off the root) and the 500,000 that receive a "down" prediction (the right branch). Depending on the outcome on day one, one of these two branches is terminated (the ones who received an incorrect prediction, who are never contacted again), and the other branch continues to be subdivided each day until we reach the final victims. Thus in hindsight the scam traverses exactly one path through a large tree, which is determined by the daily stock movements.

The pitfalls of flawed statistical reasoning due to scale and adaptivity are not as uncommon as one might think or hope. Hedge funds collectively, if perhaps inadvertently, engage in reporting practices that are

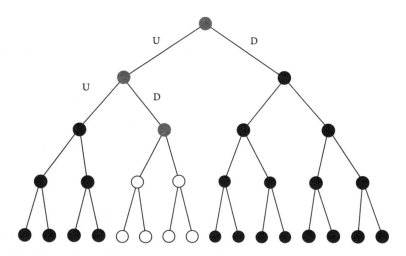

Fig. 22. Tree of possible outcomes in stock market email scam. Each level of the tree represents a day. The path of gray vertices represents the true movement of LYFT so far. The leaves of the tree represent the targets of the scam. The black leaves have already received incorrect predictions. The white leaves have received perfect predictions so far.

broadly similar to the sender's strategy in our email scam. Compared to investment banks and mutual funds, hedge funds in the United States are only very lightly regulated. In particular, they are not required to report their performance to any government authority or public database—but they can if they choose to. Voluntarily reporting their performance can be useful for attracting new investors—when the performance looks good. But when a fund is underperforming, it has a strong incentive not to report its performance. The result is that publicly reported hedge fund performance tends to overstate the returns of the broader universe of hedge funds because—just like in our email scam—the reporting is selective. (A 2012 study suggests that voluntarily reporting hedge funds may have a bias that overestimates returns of the overall universe of hedge funds by as much as 60 percent.)

And these pitfalls are not just limited to email scams, hedge funds, and other for-profit ventures. In fact, as we shall see in this chapter, the problems pervade much of modern scientific research as well.

POWER POSES, PRIMING, AND PINOT NOIR

If p-values and hedge funds are foreign to you, you have probably at least received an email forwarded from a gullible friend, or seen a post on your social media feeds, proclaiming the newest scientific finding that will change your life forever. Do you want to live longer? Drink more red wine (or maybe less). Eat more chocolate (or maybe less). Seek out pomegranates, green tea, quinoa, açai berries, or the latest superfood.

What if you want to boost your confidence before your next job interview? One of those social media posts you clicked on may have linked to a now famous 2012 TED Talk by Amy Cuddy, called "Your Body Language May Shape Who You Are", which has been viewed more than 50 million times. In the video, Cuddy proposes that spending two minutes in a "power pose"—an example is "The Wonder Woman," in which you put your hands on your hips and chin in the air—results not just in feelings of confidence but in measurable physiological changes including increased testosterone and reduced cortisol. The message is compelling: she says that "two minutes lead to these hormonal changes that configure your brain to be…assertive, confident, and comfortable."

And maybe it didn't surprise you that such a small thing could have a profound effect on your brain. You have probably been reading about

Fig. 23. Amy Cuddy demonstrating a power pose in front of Wonder Woman at PopTech 2011. Source: Wikipedia.

claims like this for years. For example, the first famous priming study conducted in 1996 showed that if you read words associated with the elderly, such as *wrinkles, Florida,* or *bingo,* you would subsequently walk more slowly. More such studies followed. Seeing the American flag would shift you toward Republican political positions. Seeing a McDonald's logo on a screen—even for such a short time that you couldn't recognize it—can make you more impatient.

These claims might seem implausible at first blush, but each is backed up by published scientific studies. So you should believe them. Right?

SCIENTIFIC GAMING

The flawed reasoning that allows our email scam to fool its victims also plagues the scientific literature. In a 2005 essay, John Ioannidis, a professor of medicine and statistics at Stanford, claimed (in the title of a paper!) that "most published research findings are false." We'll explore a couple of the causes of this problem—which are exacerbated by the algorithmization of scientific discovery—before we once again turn to algorithmic solutions.

As with our email scam, the recipe for false discovery in the sciences combines scale and adaptivity. The scale comes simply from the quantity of research that is conducted over and over again on the same datasets. The problem we are discussing is very general and afflicts medical research just as much as it affects machine learning. But for concreteness, let's think of the problem of determining whether a diet rich in açai berries makes a mouse live longer. As a rough metaphor, you can think of testing a "superfood" that actually doesn't do anything as the process of flipping a coin repeatedly. On average, you don't expect your mice to live any longer—but in your experiment they might, just by chance. In this metaphor, you expect after flipping ten coins to get five heads, but you might happen to observe more heads by chance on a particular sequence of coin flips. If enough of these coin flips come up heads—if, by chance, sufficiently many more mice live to old age than your baseline expectation—then it will misleadingly appear that you have found a promising new superfood. But of course, if you rerun the experiment with a new batch of mice—in the metaphor, if you go and flip the same coin ten more times—there is no reason to now expect that it will have any effect on the mice.

Fortunately, if you perform a sequence of ten coin flips in a row, you are very unlikely to see ten heads in a row—this is why we can generally take past performance to be a good predictor of future results, despite the small risk of being misled by unlikely variation. But as we saw in the email scam example, if you repeat the experiment a million times, you expect to see this rare event almost a thousand times! This isn't a problem on its own if the outcomes of all of the experiments are made available: the rare streaks of ten heads in a row don't make you suspect that a coin is biased if you see them in context, mixed among a sea of ordinary random-looking sequences. But the existence of lots of different experiments can become a problem if the results are only selectively shared. This is the adaptivity part.

Repeatedly performing the same experiment, or repeatedly running different statistical tests on the same dataset, but then only reporting the most interesting results is known as *p*-hacking. It is a technique that scientists can use (deliberately or unconsciously) to try to get their results to appear more significant (remember from the beginning of the chapter that *p*-values are a commonly used measure of statistical significance). It isn't a statistically valid practice, but it is incentivized by the structure of modern scientific publishing. This is because not all scientific journals are created equal: like most other things in life, some are viewed as conferring a higher degree of status than others, and researchers want to publish in these better journals. Because they are more prestigious, papers in these journals will reflect better on the researcher when it comes time to get a job or to be promoted. At the same time, the prestigious journals want to maintain their high status: they want to publish the papers that contain the most interesting and surprising results and that will accumulate the most citations. These tend not to be negative results. If goji berries improve your marathon time, the world will want to know! If goji berries have no effect on your athletic performance, there won't be any headlines written about it.

This is a game—in the game-theoretic sense we discussed in Chapter 3—and in its equilibrium the prestigious journals become highly selective, rejecting most papers. In this game, researchers have a huge incentive to find results that appear to be statistically significant. At the same time, researchers don't bother investing time and effort into projects that they don't believe have a shot at appearing in one of the prestigious journals. And negative results—for example, reports of treatments that didn't work—are not the sorts of things that will get published in high-prestige venues. The result is that even in the absence of explicit p-hacking by individual researchers or teams, published papers represent an extremely skewed subset of the research that has been performed in aggregate. We see reports of surprising findings that defy common sense—but we don't see reports of experiments that went exactly the way you would have guessed. This makes it hard to judge whether a reported finding is due to a real discovery or just dumb luck.

And note that this effect doesn't require any bad behavior on the part of the individual scientists, who might all be following proper statistical hygiene. We don't need one scientist running a thousand experiments and only misleadingly reporting the results from one of them, because the same thing happens if a thousand scientists each run only one experiment (each in good faith), but only the one with the most surprising result ends up being published.

THE SPORT OF MACHINE LEARNING

The dangers of p-hacking are by no means limited to the traditional sciences: they extend to machine learning as well. Sandy Pentland, a professor at MIT, was quoted in *The Economist* as saying that "according to some estimates, three-quarters of published scientific papers in the field of machine learning are bunk." To see a particularly egregious example, let's go back to 2015, when the market for machine learning talent was heating up. The techniques of deep learning had recently reemerged from relative obscurity (its previous incarnation was called backpropagation in neural networks, which we discussed in the introduction), delivering impressive results in computer vision and image recognition. But there weren't yet very many experts who were good at training these algorithms—which was still more of a black art, or perhaps an artisanal craft, than a science. The result was that deep learning experts were commanding salaries and signing bonuses once reserved for Wall Street. But money alone wasn't enough to recruit talent—top researchers want to work where other top researchers are—so it was important for AI labs that wanted to recruit premium talent to be viewed as places that were already on the cutting edge. In the United States, this included research labs at companies such as Google and Facebook.

One way to do this was to beat the big players in a high-profile competition. The ImageNet competition was perfect—focused on exactly the kind of vision task for which deep learning was making headlines. The contest required each team's computer program to classify the objects in images into a thousand different and highly specific categories, including "frilled lizard," "banded gecko," "oscilloscope," and "reflex camera." Each team could train their algorithm on a set of 1.5 million images that the competition organizers made available to all participants. The training images came with labels, so that the learning algorithms could be told what kind of object was in each image. Such competitions have proliferated in recent years; the Netflix competition, which we have mentioned a couple of times already, was an early example. Commercial platforms such as Kaggle (which now, in fact, hosts the ImageNet competition) offer datasets and competitions—some offering awards of $100,000 for winning teams—for thousands of diverse, complex prediction problems. Machine learning has truly become a competitive sport.

It wouldn't make sense to score ImageNet competitors based on how well they classified the training images—after all, an algorithm could have simply memorized the labels for the training set, without learning any generalizable rule for classifying images. Instead, the right way to evaluate the competitors is to see how well their models classify new images that they have never seen before. The ImageNet competition reserved 100,000 "validation" images for this purpose. But the competition organizers also wanted to give participants a way to

15 Active Competitions

2σ TWO SIGMA	**Two Sigma: Using News to Predict Stock Movements** Use news analytics to predict stock price performance Featured · 3 months to go · ❧ news agencies, time series, finance, money	$100,000 645 teams	
TGS ↗	**TGS Salt Identification Challenge** Segment salt deposits beneath the Earth's surface Featured · 9 days to go · ❧ geology, image data	$100,000 3,116 teams	
	Airbus Ship Detection Challenge Find ships on satellite images as quickly as possible Featured · a month to go · ❧ image data, object detection, object segmentation	$60,000 100 teams	
R	**Google Analytics Customer Revenue Prediction** Predict how much GStore customers will spend Featured · a month to go · ❧ regression, tabular data	$45,000 2,244 teams	
	Human Protein Atlas Image Classification Classify subcellular protein patterns in human cells Featured · 3 months to go · ❧ classification, image data	$37,000 156 teams	
RSNA Radiological Society of North America	**RSNA Pneumonia Detection Challenge** Can you build an algorithm that automatically detects potential pneumonia cases? Featured · 14 days to go · ❧ image data, medicine	$30,000 1,208 teams	
LSST	**PLAsTiCC Astronomical Classification** Can you help make sense of the Universe? Featured · 2 months to go · ❧ astronomy, time series, tabular data	$25,000 173 teams	
QUICK, DRAW!	**Quick, Draw! Doodle Recognition Challenge** How accurately can you identify a doodle? Featured · 2 months to go · ❧ writing, image data	$25,000 214 teams	

Fig. 24. Partial list of machine learning competitions hosted on the commercial Kaggle platform, many of which offer lucrative prizes to winning teams.

see how well they were doing. So they allowed each team to test their progress by submitting their current model and being told how frequently it correctly classified the validation images. The competition organizers knew this was dangerous—it risked leaking information about the validation set—but to mitigate this risk, the competition limited each team to checking their models at most two times per week.

One of the competitors, eager to be seen as a leading player in AI, was the Chinese search engine giant Baidu. Midway through the competition, Baidu announced that it had developed a new image recognition technology that put it ahead of its more established competitors, such as Google. The Baidu scientist leading the competition effort said at the time, "Our company is now leading the race in computer intelligence.... We have great power in our hands—much greater than our competitors."

But it eventually came out that the Baidu team had cheated. They had created more than thirty fake accounts to circumvent the competition rule that they could validate their models only two times per week. They had actually submitted more than two hundred tests in all, including forty tests in the five days between March 15 and March 19, 2015. By doing so, they were able to test a sequence of slightly different models that gradually fit the validation set better and better—seeming to steadily improve their accuracy. But because of how they had cheated, it was impossible to tell whether they were making real scientific progress or if they were just exploiting a loophole to exfiltrate information about the supposedly "held out" validation set.

Once the cheating was discovered, the competition organizers banned Baidu from the ImageNet competition for a year, and the company withdrew the scientific paper in which it had reported on its results. The team leader was fired, and instead of staging a coup in the competition for AI talent, Baidu suffered an embarrassing hit to its reputation in the machine learning community. But why, exactly, was testing too many models cheating? How did creating fake accounts help Baidu appear to have a better learning algorithm than it actually did?

BONFERRONI AND BAIDU

Since at least the 1950s there have been methods to curtail the dangers of running many experiments but only reporting the "interesting" ones—sometimes called the multiple comparisons problem. Here's one simple idea: if some event (say, seeing twenty heads in a row when we flip a coin twenty times) occurs only with some small probability p (say, one in a million) in a single experiment, then if we perform k such experiments, the probability that the event occurs in at least one of them will always be at most $k \times p$. Here we're just adding up the probability p in each of the k experiments. If, in our example, we perform $k = 1$ million experiments, $k \times p$ will be 1.0, warning us that seeing twenty heads in a row at least once may actually be pretty likely. So one solution to the multiple comparisons problem is that we should not report how unlikely the results would be if we performed the experiment only once, but should instead multiply this number by k. After all, if you are reporting some event, then (absent fraud) it must have happened at least once during your k tries.

This method is called the Bonferroni correction, after Italian mathematician Carlo Emilio Bonferroni. It is conservative, in the sense that if $k \times p$ is small, you are relatively safe in assuming statistical significance, whereas if $k \times p$ is large (as in the example just given), your results still might be significant, but the correction warns you against assuming so. The method is implicitly assuming the researcher selected the most misleading result—but it is "safe" in the sense that it allows for statistically valid inferences.[1]

If we go through the math and apply a Bonferroni correction to Baidu's claim of achieving higher accuracy than Google in the ImageNet competition—corrected for the number of models the company

---

[1] This might not be so far from reality: a young researcher would certainly want to present the most apparently accurate method he discovered, which (if none of them were any good) is also the most misleading one.

Fig. 25. The kind of spurious finding that could be avoided by a Bonferroni correction. Credit: XKCD, https://xkcd.com/882.

actually submitted—we still have strong enough evidence to seemingly confirm its claim. So what was the problem? It's that this methodology comes with an important caveat—the Bonferroni correction works only if the hypotheses we are testing (in this case, whether or not particular models improve in accuracy over Google's baseline) were selected in advance, before looking at the data—and it can fail us in disastrous ways if this isn't the case.

THE DANGERS OF ADAPTIVITY

The Bonferroni correction assumes that the models that are being tested are chosen before seeing the data, as opposed to the models being chosen adaptively, in response to the data. It's this latter case that presents exponentially greater methodological dangers (in a precise technical sense)—and is also the standard practice in machine learning.

Here is an illustration of how adaptation can badly lead us astray. Suppose we want to train a machine learning algorithm to predict what sorts of readers are likely to buy a copy of *The Ethical Algorithm*. Our dataset consists of a collection of one thousand historical records of people who were given the opportunity to buy the book. Some of them did, and some of them (unfortunately mired in ignorance) did not. These decisions are recorded in the data as the labels. Each record also contains various information about the individual in question—features that, for simplicity, we will assume take on only yes or no answers. It's easy to gather lots of features to throw at a machine learning problem, so let's imagine that we have done so—and it's likely that many of them won't be very useful. For example, one feature might record whether or not the potential customer's car has its fuel door on the left or the right. Another might be whether or not the person's birthday falls between January and June. Another might be whether the number of letters in the person's last name is even or odd. Our dataset could contain thousands of such features.

Remember that these apparently meaningless features need not be predictive of the labels—in fact, what we are getting at is what might happen if they are not. Let's imagine that *all* of these features are equally likely to take on values of yes or no at random, and are all completely uncorrelated with each other. And let's assume the same thing about the labels: people decide to buy our book or not by flipping a coin. (Though the assumption is unrealistic, we'd be happy with the sales figures that would result.) Thus in reality, the features have *no predictive value at all*. If we find a classifier that *appears* to have predictive performance that is better than a coin flip, we are misleading ourselves.

How should we start analyzing this dataset? One natural thing we might do is simply check whether each feature is correlated with book-purchasing behavior or not. For example, we can check how likely it is that a customer will buy *The Ethical Algorithm* if the number of letters in her last name is odd. Separately, we can check how likely it is that the customer will make a purchase if her birthday falls in the first half of the year, and so on.

Since we have a thousand people in our dataset, we expect that each of these simple ways of slicing the data will contain roughly five hundred people, and roughly half of them will have bought the book just by chance. But of course, when we flip 500 coins, we usually won't get exactly 250 heads. Some of the features in our dataset will turn out to be mildly correlated with book purchasing decisions in our data—for example, perhaps among people with an odd number of letters in their last name, 273 bought the book. Other features will turn out to be mildly anti-correlated with book purchasing decisions—maybe slightly less than half of the people with birthdays in the first part of the year made the right purchasing choice. Continuing in this way, we could note for each and every one of the features whether it was correlated or anti-correlated with the label.

What should we do with this correlation information? A very natural thing would be to combine it into the following model: for each customer, we will count how many of her features take values that are positively associated with book purchases and how many take values that are negatively associated. If there are more positively associated values than negatively associated ones, we'll predict that she will buy the book. Otherwise we'll predict she won't.

And this is where the adaptivity sneaks in. Because whether our model will count a particular feature value on the "purchase" or the "no purchase" side of the equation depends on the questions we asked about our data set: specifically, the correlations we measured in the same data set, one for each feature. This very natural idea is actually not so far from what real machine learning algorithms—with names like "bagging" and "boosting"—do to combine weak correlations into powerful predictive models.

Unfortunately, this sensible-sounding methodology can quickly lead us astray. If you try this out with enough features, you'll find that you appear to get a classifier that approaches perfect accuracy—able to predict with near certainty who will buy the book—if you measure the performance of your classifier on the same data set you trained it on. Of course, we know that on new customer data, this classifier won't do better than random guessing—because customers are just flipping coins. What's worse is that if you apply a Bonferroni correction to the number of questions you actually asked—the questions "Is this feature correlated with the label or not?" for each feature in the data—it will still seem to confirm the impressive accuracy of your classifier.

THE PATHS NOT TAKEN

The reason this happens is that when constructing our classification rule, we used information we "peeked" from the dataset by asking about the individual feature correlations. Even though none of the features we tested were truly correlated or anti-correlated with the label we were trying to predict, we learned whether they happened to be slightly associated with the label in our particular dataset. So if we took a random person in our dataset and a feature that by chance was slightly correlated with book buying, the purchasing behavior of this person would be slightly more likely than not to agree with our esoteric feature—but only because we found it to be correlated with the label in the same data that we are testing the classifier on. Combining many of these features together compounds the small advantage each feature has many times, eventually leading to a perfect-looking classifier—although ultimately it is doing nothing other than fitting the noise in the data. And note that the procedure that we described hardly seems malicious. In fact, it seems like an extremely reasonable way to learn from data: first find features that are individually weakly predictive of the label, and then combine them in a sensible way. This is why it is important in a machine learning competition to limit competitors from being able to repeatedly query the validation data, and why it is important in empirical science not to allow ourselves too much access to the data when formulating our hypotheses—even when we believe that we are behaving reasonably. It can be easy to p-hack without intending to.

The problem in this example is actually very closely related to the email scam that we began the chapter with. To understand what is going on, consider the decision tree diagram in Figure 26, which you may remember from our discussion of the email scam. The circles are called "vertices," and the one at the top is called the "root" of the tree. The ones at the bottom are called the "leaves." (Yes, this is upside down: computer scientists apparently don't get out of the office enough to have seen real trees.)

Imagine that every level of the tree corresponds to one of the features in our dataset. For each feature, we have asked whether it is correlated or anti-correlated with the label. If it is correlated, we take the left branch of the tree. If it is anti-correlated, we take the right branch. Finally, when we get to the end of a path—a leaf of the tree—we have gathered enough information to construct a classifier, which is a majority vote to see if an example has more correlated or anti-correlated feature values in it. Since every leaf corresponds to a different path

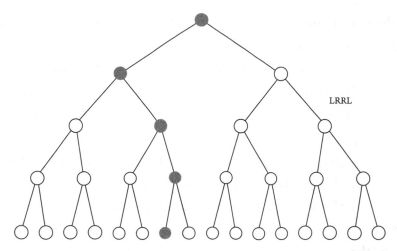

LRRL

Fig. 26. Tree illustrating the dangers of adaptive data analysis and *p*-hacking. Each level of the tree corresponds to a feature that could be correlated (left) or anti-correlated (right) with the label. The gray path (LRRL) represents the outcomes of the correlation tests. Each leaf corresponds to a classifier that results from a sequence of correlation tests.

down the tree, each leaf corresponds to a different classification algorithm we might have arrived at.

If there are d features in our dataset, then a path down the tree corresponds to having asked d questions—the shaded vertices in the figure. But the number of classifiers we might have arrived at—the total number of leaves—is vastly larger than this: it is 2^d. If $d = 30$, then we are already choosing among more than a billion classifiers. More generally, here is a way of counting the classifiers we might have arrived at, which will be useful later: we can label each leaf of the tree by the sequence of left or right decisions we had to make to traverse our way down the tree to get to it. If the first feature we tested was correlated with the label, we go left. If the second feature we tested was anti-correlated with the label, we go right. We arrive at each leaf through a unique sequence of d of these left/right choices, which we can label with this sequence—for example, LRRL... Since there are two choices in each of d positions, there are 2^d different leaves.

The reason the Bonferroni correction doesn't work is that it is not enough just to correct for only the d questions that we actually asked—the shaded path down the tree. We must also correct for the 2^d questions we *might* have asked had the answers to our earlier questions come out differently. That is, we have to correct for all of the leaves in the tree, not just the path we actually took. As this example shows, it is possible we have to correct for an exponentially larger set of models than we explicitly evaluated. This is the same phenomenon that fooled the victim of our email scam: he should have taken into account all one million initial targets of the scammer, rather than assuming that he was the only one involved.

IF YOU TORTURE THE DATA

Of course, it is generally impossible to correct for all questions that might have been asked. In our simple example, we can explicitly imagine the decision process that we went through as we tested each variable in turn, and count the number of different results that we might have arrived at. But in a more typical situation, sequential decisions are being made at least in part by human beings, and it isn't possible to reason about the myriad counterfactuals: *What would I have done in every circumstance had my analyses come out differently?* This is why making decisions based on the data has classically been viewed with skepticism, going by various derogatory names including "data snooping," "data dredging," and—as we have seen—"p-hacking."

The methodological dangers presented by the combination of algorithmic and human p-hacking have generated acrimonious controversies and hand-wringing over scientific findings that don't reflect reality. These play a central role in what is broadly referred to as the "reproducibility crisis" in science, which has its own Wikipedia pages that begins:

The **replication crisis** (or **replicability crisis** or **reproducibility crisis**) is an ongoing (2019) methodological crisis in science in which scholars have found that the results of many scientific studies are difficult or impossible to replicate or reproduce on subsequent investigation, either by independent researchers or by the original researchers themselves. The crisis has long-standing roots; the phrase was coined in the early 2010s as part of a growing awareness of the problem.

While *p*-hacking is not the only culprit here—poor study design, sloppy experimental technique, and even occasionally outright fraud and deception are present—the concern is that even well-intentioned data-driven scientific exploration can lead to false findings that cannot be reproduced on fresh data or in fresh experiments. Among the more prominent examples associated with *p*-hacking was the controversial research on "power poses" that we discussed earlier. Here is how it was described in the *New York Times Magazine* in late 2017:

> The study found that subjects who were directed to stand or sit in certain positions—legs astride, or feet up on a desk—reported stronger "feelings of power" after posing than they did before. Even more compelling than that, to many of her peers, was that the research measured actual physiological change as a result of the poses: The subjects' testosterone levels went up, and their cortisol levels, which are associated with stress, went down.

But the study failed to be replicated, and "power poses" became the poster child for the reproducibility crisis and the dangers of *p*-hacking.

> "We realized entire literatures could be false positives," [prominent *p*-hacking critic Joe] Simmons says. "They had collaborated with enough other researchers to recognize that

the practice was widespread and counted themselves among the guilty."

The famous priming studies we discussed also failed to hold up to scrutiny. And food science has been under suspicion for years. A notable *p*-hacking scandal rocked the food science community in 2017. In this instance the principal researcher in question, a celebrated Cornell professor named Brian Wansink, seemed to actively embrace *p*-hacking as a means of generating results. The following is a description of instructions he sent to a student working with him, whom he wanted to find something interesting about all-you-can-eat buffets:

First, he wrote, she should break up the diners into all kinds of groups: "males, females, lunch goers, dinner goers, people sitting alone, people eating with groups of 2, people eating in groups of 2+, people who order alcohol, people who order soft drinks, people who sit close to buffet, people who sit far away, and so on …"

Then she should dig for statistical relationships between those groups and the rest of the data: "# pieces of pizza, # trips, fill level of plate, did they get dessert, did they order a drink, and so on …"

"This is really important to try and find as many things here as possible before you come," Wansink wrote to [the student]. Doing so would not only help her impress the lab, he said, but "it would be the highest likelihood of you getting something publishable out of your visit."

He concluded on an encouraging note: "Work hard, squeeze some blood out of this rock, and we'll see you soon."

[The student] was game. "I will try to dig out the data in the way you described."[2]

2 From *Buzzfeed News*, February 2018.

Here we have a research methodology virtually encapsulating our tree of spurious correlations: create a very large space of features (gender, lunch or dinner diners, size of group, location in the restaurant, etc.), and then go on a fishing expedition by making queries on that data. After these practices were revealed, intense scrutiny of Wansink's research led to retractions of seventeen published papers, another fifteen "corrections," and a Cornell investigation that found scientific misconduct. He resigned from the university effective June 2019.

But these cases are really just an acceleration of an old phenomenon. As Ronald Coase, a Nobel Prize-winning British economist, put it in the 1960s, "If you torture the data for long enough, it will confess to anything."

TENDING THE GARDEN OF THE FORKING PATHS

Andrew Gelman and Eric Loken, statisticians who have studied the proliferation of published but false findings in the social sciences, have a colorful name for the phenomenon of adaptivity: the "garden of the forking paths." The forking that Gelman and Loken are referring to is exactly the branching we saw in the tree illustration. The tree diagram itself is a map of the garden of choices that we committed ourselves to once we decided on our analysis procedure. The term is intended to indicate that the kind of overfitting we saw that can result from adaptive questioning need not be intentional or malicious: an earnest scientist can mislead himself by getting lost in the garden.

But unintentional or not, false discovery is harmful and expensive. A 2015 study estimated that the monetary cost of irreproducible preclinical medical research exceeds $28 billion per year. This monetary cost is on top of the scientific cost, ultimately slowing the search for lifesaving cures by wasting the time of medical researchers and giving false hope to patients waiting for breakthroughs that could save their lives. Getting lost in the garden of the forking paths is an ethical issue, not just an academic one. Data analysis is vastly increasing in scale and complexity, so we need to look for algorithmic solutions.

Our tree diagram was a map of the garden that results from our particular machine learning procedure. But for every different procedure we might run, there is a different tree—a different garden. And if the procedure is complicated or imprecisely specified—for example, if a human being or an entire research community is involved anywhere in the decision-making process—we won't have the luxury of having a precise description of the tree, or a garden map. What is needed are methods and algorithms that can limit the risk of false discovery even in cases when we do not have a map.

One safe but extreme measure that has recently gained popularity in parts of the social sciences is study preregistration. As described in *The Preregistration Revolution,* "Pre-registration of an analysis plan is committing to analytic steps without advance knowledge of the research outcomes." In other words, the goal of preregistration is to prevent the data analyst from being able to make any decisions at all during the analysis. It does this by forcing the scientist to publicly commit to a detailed analysis plan before having a chance to look at the data. When implemented carefully and correctly, preregistration removes the risk of the garden of the forking paths by gating it off, preventing the researcher from ever entering it. But faithfully following a preregistration plan means creating it *without any influence at all* from the data. Ideally it is registered before the data is even gathered.

While safe (and a big improvement over unprincipled, unfettered data analysis), preregistration is highly conservative. Scientific insight doesn't generally arise from nothing, as happened with Archimedes and the flash of insight in the old story.[3] Instead, scientific progress is achieved as men and women stand on the shoulders of giants: the

---

[3] According to myth, Archimedes discovered that an object displaces a volume of water equal to its own volume when he stepped into a bath, and proclaimed, "Eureka!"

1. Create a registration

Open the project that you want to register, then click the **Registrations** tab in the navigation bar.

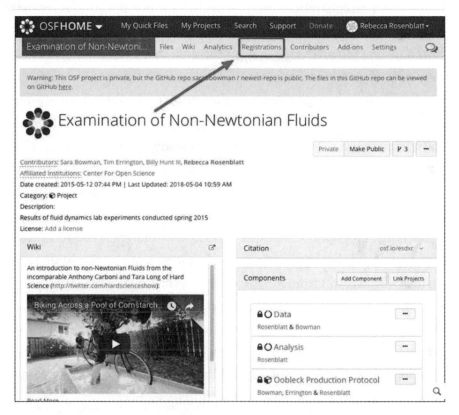

Fig. 27. Project and data registration interface on the website of the Open Science Foundation, one of the most prominent organizations providing preregistration of datasets, hypotheses, and analyses. To further guard against *p*-hacking, preregistrations cannot be deleted.

accumulated work of their predecessors. What this means in practice is that scientists read each other's papers and get inspiration from the findings of their peers. When they then go on to test their own ideas, they cannot always gather fresh data, because large, high-quality datasets are rare and expensive. This is the case in machine learning, in which datasets such as ImageNet contain millions of human-labeled images, and is one of the primary reasons that the research community relies on a small number of benchmark datasets. Certain kinds of medical datasets can be even more difficult to gather. And every time a dataset is reused, a complicated fork is created in the garden path that even preregistration can't prevent: simply because scientists have been exposed to previous research ideas, the complicated decision-making processes they go through to take a new step creates new data-dependent decisions. To correctly adhere to preregistration, every study would need to gather its own dataset, which is often unrealistic.

Fortunately, a recent set of algorithmic advances points to a solution: a methodology to safely reuse data, so that we avoid false discoveries while continuing to be guided by the data. These methods allow us to be inspired by data—to explore the garden of the forking paths without a map or, in other words, without the need to understand all of the delicate contingencies that an analysis might depend on and which result from human decision-making. Instead they work by directly constraining how the garden can be built, by sitting as an algorithmic intermediary between the data analyst and the data. We are just beginning to understand how these new tools should work—even in theory. In the rest of this chapter, we'll explore the algorithmic ideas that underlie these nascent methods.

THE GARDEN'S GATEKEEPER

The garden forks are constructed when a "data analysis procedure"—which might be a machine learning algorithm, a human being, or a team of researchers—interacts with the data. If we want to allow human researchers the freedom to make decisions as they like, we have no control over the data analysis procedure. But we can control *how* the procedure—whether an algorithm or a human—interacts with the data.

For example, the organizers of the ImageNet competition already limited how competitors could interact with the validation set: all teams could do was submit their candidate machine learning models and then see how accurately they classified the images in the validation set. They couldn't get direct access to the actual images in the dataset. As we saw, this wasn't enough of a limitation to prevent incorrect findings—but it provides us a good starting point to imagine how we might mediate access to data. Data analysts will submit questions about the data to an algorithmic interface that we get to design, and the goal of this algorithm will be to provide accurate answers to those questions. Crucially, we will want the answers to be accurate with respect to hypothetical new data drawn from the same source as our dataset, not just on the dataset that we have. This way, we won't be graded as "accurate" if we simply fit the noisy peculiarities of our particular dataset.

Let's now stretch our imaginations a little and ask what would happen if our interface had a property that at first seems implausible (but that we will shortly see can be made quite practical in certain cases). Suppose that we made the promise that no matter what sequence of questions the data analyst decided to ask to our interface, the sequence of answers it provided could always be summarized by just a small amount of information—say by a short string of just k zeros and ones, where k is some number that is much smaller than the number of questions asked. Would this property let us say something about how a data analysis procedure might traverse the garden of forking paths?

Let's think about how a data analysis procedure (call it P) would interact with this interface. P can represent either an algorithm or a human being, or some combination of the two—it doesn't matter. Just as in our machine learning thought experiment, P will proceed by asking questions and getting answers. First, P will ask question 1. The interface will provide some answer to P's question, and depending on what the answer is, P will pick some question 2 to ask. The interface provides an answer to P's second question, and depending on what the answer is, P asks a third question, and so on. The procedure is adaptive because the choice of which question P asks at each step depends on what answers it received in the previous steps.

Once we decide on a particular data analysis procedure P, in principle we have everything we need to map out a garden of forking paths, like we did with our tree diagram, because we could work out what question would be asked at each step, in every possible eventuality. Of course, this is possible only in principle, because for every P there will be a different garden, and if P involves any human decision-making, we won't know a precise enough description of it to actually figure out what would happen in each eventuality. Even if we do know a precise specification (because P is an algorithm), it will almost always be far too complicated to reason about (let alone map out) the tree of all possible contingencies. This is the problem we need to confront.

But although we cannot say anything about the tree of contingencies that P builds without knowing the details of P, if we know that the sequence of answers produced by the interface can always be summarized by a short sequence of k characters, then we can say quite a bit about the shape of the garden. Remember back to our discussion of the tree diagram: we could name each leaf of the tree by the sequence of left and right decisions we had to take to get there—or equivalently, a sequence of d zeros and ones. If we asked d questions in total, this label would be a sequence of length d. But now we can label each leaf that is actually ever reached by a sequence that has length just k—because we must have been guided to that leaf by one of the possible sequences of answers that the interface can produce, and we know that all of these can be described with just k characters. So if k is much smaller than d, we know that it must be that we can't actually ever arrive at most of the leaves of the tree—because there are far fewer sequences of length k than there are sequences of length d. The set of questions that we can actually ever ask is much smaller than we thought—and we can apply a Bonferroni correction to this much smaller set of questions that our procedure might ask. We can do this even though we don't know which questions these are—it's enough to know that the set of possible questions is small.

Of course, the harder question remains: how could an interface give useful answers to d questions in a way that can be summarized by a description that is much smaller than d?

A SUCCINCT LEADERBOARD

In the ImageNet competition, the point of giving competitors access to an interface that let them test their model on the validation data was to give them a sense of their standing in the competition. But a competition is a black-and-white affair: there are winners and losers. If your model is not the best one that has been submitted so far, it's not clear that you need to know exactly how much worse your model is compared to its competitors. It might be enough for you to know that you are not yet in the lead.

Consider an interface that takes as input a sequence of machine learning models, submitted by the ImageNet competitors. For each model, it tests whether its classification error is at least 1 percent lower than the best model that has been submitted so far. If the answer is yes, the interface says so, and answers with an estimate of the classification error of the submitted model, which is now the algorithm to beat. But if the answer is no, it doesn't provide any estimate of the classification error of the model—it simply reports that the submitted model does not (substantially) improve on the previous state of the art.

It turns out that the answers provided by this interface always can be summarized concisely. Here is why. Suppose that over the course of the competition, the competitors submit 1 million models to be validated. Even though this is a huge number, we know that at most 100 of these models could have improved by 1 percent or more over the best previous model—and probably far fewer. This is because error rates are numbers between 100 percent (getting everything wrong) and 0 percent (getting everything right). If your model improves by at least 1 percent over the best previous model, it makes the state-of-the-art error smaller by at least 1 percent. And you can't decrease a number that starts at 100 by at least 1 more than 100 times before you get down to zero. In a typical case, the number of improvements will be much smaller than 100, because the first model won't get everything wrong, the best model won't get everything right, and some models will improve by more than 1 percent on the best result so far.

So all we need in order to write down the sequence of answers is which 100 of the 1 million models submitted were the ones that improved, and the reported error of those 100 models. One way to do this is just to write down a list. Each entry in the list records the index of one of the models that yielded an improvement of at least 1 percent, together with its error. For example, the list might look like this: (1, 45%), (830, 24%), (2,817, 23%), (56,500, 22%), (676,001, 15%). What this means is that the models that improved over the previous best were the 1st one submitted, the 830th one submitted, the 2,817th one, the 56,500th one, and the 676,001st one. The errors of those models were 45, 24, 23, 22, and 15 percent, respectively. This is enough to reconstruct the entire output of the interface, because we know that at every other round, the interface must have reported no improvement. It is possible to summarize the set of answers concisely because we know that no matter what, this list will be short—it can't have more than 100 entries. The result is that we can guarantee that the answers to these queries will be almost as accurate as they would have been had there been no adaptivity at all (e.g., if all of the models had been preregistered). This effectively eliminates the possibility either that a data analyst will get lost in the garden of the forking paths and inadvertently trick himself or that a competitor (like Baidu) trying to cheat will be able to.

This idea solves the problem of maintaining a "leaderboard" to keep track of who is currently winning a machine learning competition. But what about if we want a more fully functional data interface, which can provide answers to a richer set of questions? By applying the right tools from machine learning combined with the ideas we just discussed, it turns out that we can answer an arbitrary sequence of questions with error that is comparable to what we could guarantee had the study been preregistered. By using a clever algorithm, we avoid needing to wall off the data analyst from the garden.

TENDING A PRIVATE GARDEN

As it turns out, differential privacy, the tool for private data analysis that we saw in Chapter 1, can lead to further improvements beyond the methods we just discussed. A recent discovery—one that seemed surprising at the time, but in retrospect is perhaps natural—is that algorithms that are differentially private cannot overfit. What this means is that if we can design a differentially private algorithm that is able to accurately answer questions with respect to a fixed benchmark dataset, then those answers will also faithfully reflect the answers to our questions evaluated on new data, so long as it comes from the same source (distribution) as our benchmark dataset. This is good— it allows us to draw on more than a decade of research designing differentially private algorithms to address the problems of being misled by data analysis and machine learning. The connection is natural, because as we discussed in Chapter 1, differentially private algorithms are designed to allow researchers to extract generalizable statistical facts about a population while preventing them from learning too much about any particular individual in that population (which we also saw had useful incentive properties for game theory in Chapter 3). This turns out to align almost exactly with the goals of a researcher who wants to advance science without getting lost in the garden of the forking paths. She wants to learn generalizable facts about the world without accidentally overfitting the noisy peculiarities of her sample of data. And you can't fit the noisy peculiarities of a dataset without fitting the peculiarities of individual records in that dataset—which is exactly the thing that differential privacy prevents.

The upshot of all of this is that the problem of false discovery in scientific research can be tackled by the careful formalization of the problem and the deployment of tools from algorithm design. For this particular application, we are still in early days, and much remains to be done.

For example, the general purpose algorithms are still tremendously computationally expensive and would take too much time to run in real applications. Far more efficient algorithms will be needed before we can put this approach to work. But it already makes clear that the solution to a problem that is greatly exacerbated by algorithms—because of the scale and adaptivity with which they can explore the garden of forking paths—will itself lie in the realm of algorithm design. And it highlights how precisely formulating algorithmic goals and then designing algorithms to satisfy them is a science that itself leads to modular, generalizable insights. Here, just as in Chapter 3, we see an example in which techniques designed for one purpose—privacy—prove useful even in a setting in which privacy is not itself a goal.

Risky Business

Interpretability, Morality, and the Singularity

LOOKING FOR THE LIGHT

This book has focused on the potential for algorithms and learned models to violate basic social values, and how we might prevent these violations with better science. This is clearly a timely and important topic. But we are not immune to the criticism that we have been "looking where the light is" in the choice of the specific values or norms we have emphasized. For instance, fairness and privacy are perhaps the two areas of research on ethical algorithms that have received the most scientific attention and have the most mature literatures, theories, and experimental methodologies (nascent as they still are), so they have featured prominently here.

There are good technical reasons for this. Research in privacy, for example, has greatly benefited from the fact that scientists seem to agree on definitions that capture what is intuitively meant but are also precise and manageable. There is less agreement about what "fairness" should mean, but there are at least a number of concrete proposals. This is not to say these definitions are entirely settled (indeed, we saw in Chapter 2 that entirely reasonable definitions of fairness can be in conflict with each other), or that anywhere near all the algorithmic problems have been solved. But solid mathematical definitions are the starting point of any rich and useful theory, and privacy and fairness are off to fast starts in this regard. Similarly, our chapters on algorithmic game theory and preventing p-hacking were partly chosen for the relative maturity of the research in those fields.

But what about other values, such as having algorithms and models that are "transparent" or "interpretable"? Or algorithms that are "accountable" for their actions, or can even be said to be "safe" or "moral"? Despite the scare quotes, these are obviously all properties that we desire and strive for—and sometimes demand—from human decision-makers and organizations.

We haven't focused on these not because they aren't important but simply because there is less to say about them at this point—at least of the precise type in which we are interested, based on relatively stable definitions and algorithmic design principles. But in this final chapter we'll provide some brief discussion of norms and values that seem to have largely eluded precise algorithmic formulations, and say a bit about why this might be so. We'll proceed roughly in order of increasing generality, starting with things such as algorithmic interpretability, then moving on to more slippery topics, such as morality—and concluding with every AI alarmist's favorite dystopia, "The Singularity."

LIGHTING THE BLACK BOX

Perhaps closest in spirit to the kinds of values we have focused on so far are those of transparency and interpretability. The goals are simple to express informally: we would like our models to be things we can understand and therefore trust. The difficulty lies not in expressing this intuition but in formalizing it in a way that is both general and quantitative.

While individuals might disagree on how important fairness and privacy are, and on how much of them we should demand, they seem to broadly agree on what the things are themselves—or at least what they are not. An algorithm for college admissions decisions that has a false rejection rate for black applicants that is much higher than for white applicants seems unfair to most people. There might be debate over how much higher constitutes real harm, but not over the nature of the unfairness itself. Similarly, people may have differing opinions on how much of their personal data should be allowed to inadvertently leak, but they seem to agree that such leakages constitute violations of what we'd call privacy. In fact, the quantitative definitions of fairness and privacy that we studied here provided explicit "knobs" or parameters that let a user tell an algorithm exactly how much fairness or privacy they demand. This is part of what we mean by a precise algorithmic theory.

In contrast, the first question that comes to mind when attempting to formulate a theory of algorithmic interpretability is, interpretable to whom? The very word *interpretable* implies an observer or subjective recipient who will judge whether she can understand a model or its behaviors. And our assumptions about the universe of potential recipients and their level of numeracy could have dramatic effects on even qualitative definitions of interpretability.

Let's make this issue a bit more concrete and consider the specific question of whether the model output by a machine learning algorithm is interpretable or not. To someone with little or no quantitative education, the concept of a precise mathematical object mapping, say, features extracted from a loan application to a prediction of repayment likelihood might be completely alien. They might simply not have the training to think concretely about such a function, and to them no such model is interpretable. To an observer familiar with the basics of classical statistical modeling, but perhaps not all of modern machine learning, a linear model that takes a weighted sum of the input features would be the canonical example of interpretability—in particular, one can look at the signs of the weights and see that (for example) being in your current job for many years is positively correlated with loan repayment, and having a criminal record is negatively correlated. However, this same observer might deem a multilayer neural network not interpretable because such simple correlations cannot be easily extracted. But to a practitioner in the field of deep learning, this is not a barrier, since more complex, higher-order correlations can still explain the model's behavior.

If we take the entire scope of mathematical and computational literacy encompassed by these different observers and try to formulate good definitions of interpretability for each, we'll basically get the range from "no algorithms allowed" to "anything goes." The subjectivity involved seems to make algorithmic theories of interpretability difficult.

Difficult, but perhaps not impossible. The discussion above suggests the following broad approach to developing (perhaps multiple) definitions of interpretability:

- Identify the target population or group of observers—for example, high school graduates with no further education, employees of the regulatory agency overseeing credit scoring algorithms, and so on.
- Design and run behavioral experiments with the observers in which they are asked whether they understand algorithms and models of varying types and complexity and are asked questions testing that understanding.
- Use the results of the experiments to formulate an observer-specific definition or measure of interpretability.
- Study the design and limitations of algorithms obeying this definition.

It would seem that this agenda is a solution (perhaps the only solution?) to the inherent subjectivity of interpretability. But so far, there is relatively little research on the behavioral aspects of interpretability across different groups of observers. More often, the research skips the experimental step and simply equates interpretability with some particular class of models—for example, linear models only, or maybe linear models with coefficients of small magnitude—and then jumps to the last step of the program above. But without really identifying your target audience, it seems impossible to answer the recurring question: interpretable to whom?

You can use this table to choose a suitable interpretable model for your task (either regression (regr.) or classification (class.)):

Algorithm	Linear	Monotone	Interaction	Task
Linear models	Yes	Yes	No	Regr.
Logistic regression	No	Yes	No	Class.
Decision trees	No	Some	Yes	Class. + Regr.
RuleFit	Yes	No	Yes	Class. + Regr.
Naive Bayes	Yes	Yes	No	Class.n
k-nearest neighbours	No	No	No	Class. + Regr.

A table describing the interpretability of various types of models based only on their abstract mathematical properties, rather than the extent to which they can be understood by different groups of people. From Christoph Molnar, "Interpretable Machine Learning: A Guidebook for Making Black Box Models Explainable," 2018.

Another challenge for interpretability research—one that seems of equal importance with the issue of subjectivity—is the question of what we want to be interpretable in the first place. Consider the standard machine learning pipeline: historical data is collected by some mechanism, which is then fed to a learning algorithm that searches for a low-error model on the data, which in turn is used to make future decisions on new data "in the field." There are (at least) four distinct entities here whose interpretability we could discuss: the data, the algorithm, the model found by the algorithm, and the decisions made by the model.

For instance, we have already argued in the introduction that most common machine learning algorithms are rather simple (not too many lines of fairly straightforward code), principled (maximizing a natural and clearly stated objective, such as accuracy), and thus are in some sense already interpretable or understandable. However, the models output by such algorithms can be extremely difficult to fully understand, and they may capture complex and opaque relationships between the variables in what seemed to be a simple dataset (such as records summarizing the financial, credit, and employment history of past loan applicants, along with the outcome of their loans). Thus simple algorithms applied to simple datasets can nevertheless lead to inscrutable models. (And, of course, more complex algorithms applied to complicated datasets can lead to even greater model opacity.)

But it's also entirely possible that even a model we can't understand "in the large" can make specific decisions that we *can* understand or explain. For instance, suppose your loan application has been rejected by a deep neural network. One of the benefits of algorithmic decision-making is that we have models that tell us what they would do on any input, so we can explore counterfactuals such as "What would be the smallest change to your application that would change the decision to acceptance?," to which the answers might be things like "Be in your current job for six months longer" or "Have more equity in your home for collateral." Indeed, this type of explanatory understanding at the level of individual decisions or predictions is the basis for some of the more promising research on interpretability.

Finally, even when confronted with complex models comprising thousands or more interacting components, we can still try to glean important insights about their inner workings. For example, a deep

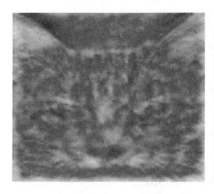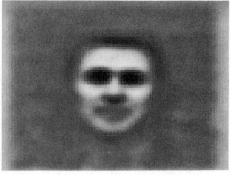

Fig. 28. Optimal stimuli for a "face neuron" and a "cat neuron" in a deep neural network trained on image data. From Google Research.

neural network has many internal "neurons" that are intended to learn higher-level properties of the data. To try to understand what any particular neuron has learned, we can ask what its optimal stimulus is—that is, the input to the network that causes it to activate or "fire" most strongly. This is directly analogous to, and inspired by, methods in neuroscience, where, for instance, studies have discovered highly specialized neurons in mammalian brains that are devoted to detecting motion in the visual field. Similar studies on neural networks trained on video and image data reveal neurons specialized for detecting human and feline faces. Of course, other neurons in the same network may have no clear function we can identify (just as in biological brains), so such techniques are rather hit-or-miss, and thus don't seem to be a promising general methodology for interpretability.

SELF-DRIVING MORALITY

As important as fairness, privacy, and interpretability are, we usually do not consider violations of them to be an immediate threat to our health and physical safety (although they can be in areas like criminal justice). But as algorithms come to play central roles in areas such as self-driving cars, personalized medicine, and automated warfare and weaponry, we are inevitably confronted with issues of algorithmic safety, morality, and accountability. While these are topics are being actively discussed in both scientific circles and the mainstream media, there is even less technical progress on them than there is on transparency. And perhaps this is as it should be—while the scientific agenda of this book has been the precise specification of social values and their subsequent algorithmic internalization, maybe there are some norms that we can't or don't want to formalize and don't want algorithms to encode or enforce. In fact, perhaps there are some kinds of decisions that we never want algorithms to make, period—even if they make them "better" than humans. We'll return to this point shortly.

Some of the popular and scientific discussion of algorithmic morality has focused on thought experiments highlighting the difficult ethical decisions that self-driving cars and other systems might soon confront on a regular basis. The Moral Machine project at MIT presents users with a series of such dilemmas in an effort to poll human perspectives on AI and machine learning morality. While it

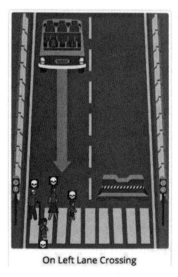
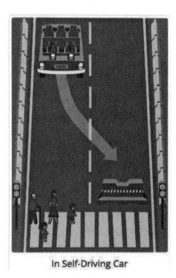

On Left Lane Crossing | In Self-Driving Car

Fig. 29. Illustration of standard hypothetical moral dilemmas faced by self-driving cars, in which the controlling algorithm must decide whether to sacrifice its passengers or the pedestrians. From the Moral Machine project at MIT.

might seem like an extended parlor game, perhaps projects such as this will eventually gather valuable subjective data on moral perception, somewhat akin to the suggestion of surveying user groups to advance algorithmic transparency.

For the most part, computer scientists are newcomers to the topics of ethics and morality; in contrast, philosophers have unquestionably devoted the longest and deepest thought to them. In fact, the dilemmas presented in the Moral Machine project are strongly reminiscent of the thought experiments with which political philosopher Michael Sandel has been challenging the students in his Harvard course on justice for many years. While Sandel has not addressed algorithmic morality specifically, he has written on topics that strike us as closely related. For instance, his 2012 book *What Money Can't Buy: The Moral Limits of Markets* discusses the ways in which creating a market for something that previously didn't have one can fundamentally alter the nature of the thing itself. Examples include paid services that will stand in line for you to get into congressional hearings or get free tickets to Shakespeare in Central Park, and the commercialization of incarceration by privately managed prisons. Sandel writes:

> The most fateful change that unfolded during the past three decades was not an increase in greed. It was the expansion of markets, and of market values, into spheres of life where they don't belong. To contend with this condition, we need to do more than inveigh against greed; we need to rethink the role that markets should play in our society. We need a public debate about what it means to keep markets in their place. To have this debate, we need to think through the moral limits of markets. We need to ask whether there are some things money should not buy.

This view is contentious among economists—but if one replaces "markets" by "algorithms," and "some things money should not buy" by "some decisions algorithms should not make," it would seem that both this passage and many of the arguments Sandel makes still apply. Maybe the fundamental nature of decisions can change when an algorithm makes them instead of a person.

What might be examples of decisions algorithms should not make, and why? In the sphere of automated warfare, some have argued that an algorithm should never make the decision to kill a human being— even if it could be demonstrated that the algorithm could make more accurate decisions (for example, in distinguishing between enemy combatants and civilians, or in killing only the target without any collateral damage). The argument is that the final decision to kill a human being should only be made by another human being because of the moral agency and responsibility involved; the weight of such a decision should lie only with an entity that can truly understand, in a human way, the consequences at hand. Of course, if the algorithm really is more accurate, sticking to this moral principle will result in the deaths of more innocent people. Readers can think about where they believe the moral limits of algorithms should lie.

But now that we've reached the point of discussing the ethical issues surrounding algorithms that can cause physical harm or even death, we may as well take such concerns to their logical extreme: algorithms as a potential existential threat to the human race. Although fanciful, we'll see that this concern is just a logical extreme of several of the themes we have seen throughout this book.

A SINGULAR THREAT

> I really hate this damn machine,
> I wish that they would sell it.
> It never does just what I want,
> but only what I tell it.
> —Anonymous

In December 2017, huge wildfires were roaring through Southern California, forcing the evacuation of tens of thousands of people, and engulfing highways around Los Angeles. People fleeing their homes naturally turned to their navigation apps, such as Waze and Google Maps, to find the fastest route to safety. As we discussed in Chapter 3, navigation apps are usually a better bet to get good directions than using an old-fashioned paper map, because the apps have lots of real-time data and powerful optimization engines. Because they record data from the other users of the app, they know a lot about current traffic conditions on various roads and can predict future traffic conditions. Then they optimize to find you the fastest route to where you are going.

But during the California fires, Waze was reportedly directing people *toward* the roads that were engulfed in flames. If you think about it, there is a certain logic to this: those roads were entirely free of traffic. But this is clearly not the outcome that Waze users wanted, or that its engineers intended.

This is a simple example of optimization gone awry, but if you let your imagination run free, you can envision much more serious problems. If you extrapolate a little from the kinds of technology we have today to the kinds we might have in a hundred years, you might even arrive at the thought that AI poses an existential threat to humanity. And many high-profile people have. Stephen Hawking said that superintelligent AI "could spell the end of the human race." Elon Musk views artificial intelligence as "our greatest existential threat." And Google DeepMind cofounder Shane Legg has said that he thinks that artificial intelligence poses "the number one risk for this century." In fact, when Google negotiated the purchase of DeepMind in 2014 for $400 million, one of the conditions of the sale was that Google would set up an AI ethics board. All of this makes for good press, but in this section, we want to consider some of the arguments that are causing an increasingly respectable minority of scientists to be seriously worried about AI risk.

Most of these fears are premised on the idea that AI research will inevitably lead to superintelligent machines in a chain reaction that will happen much faster than humanity will have time to react to. This chain reaction, once it reaches some critical point, will lead to an "intelligence explosion" that could lead to an AI "singularity." One of the earliest versions of this argument was summed up in 1965 by I. J. Good, a British mathematician who worked with Alan Turing:

> Let an ultraintelligent machine be defined as a machine that can far surpass all the intellectual activities of any man however clever. Since the design of machines is one of these intellectual activities, an ultraintelligent machine could design even better machines; there would then unquestionably be an "intelligence explosion," and the intelligence of man would be left far behind. Thus the first ultraintelligent machine is the last invention that man need ever make, provided that the machine is docile enough to tell us how to keep it under control.

We seem to be on our way to building machines that can match or surpass human ability at a variety of challenging tasks—recently, including things like chess, *Jeopardy*, Go, and perhaps even driving a car. So why shouldn't we expect that eventually we will be able to build machines that match human ability at designing algorithms for learning, optimization, and reasoning—that is, algorithms for artificial intelligence? These algorithms will be as good as we are at finding improved techniques for learning and optimization. But once an algorithm finds those improvements, it has the ability to reprogram itself, further speeding the pace of discovery. These improvements will build on themselves, snowballing.

Suppose we accept this chain reaction argument—that because the design of intelligent machines is itself something that could be better done by more intelligent machines, then once we reach some

critical level of artificial intelligence, we will have the ability to produce machines that are vastly more intelligent than human beings.
Clearly this is something that has the potential to be dangerous, at
least as a weapon. If an intelligence explosion were close at hand, we
could predict an arms race in which the United States, China, Russia,
and other countries rush to weaponize AI. (Some feel that this arms
race has already begun.) But in this scenario, the real threat is the hos-
tile foreign power, not the AI per se. The AI serves only as a dangerous
weapon, akin to (but potentially much more destructive than)
nuclear weapons.

Is there a scenario in which the threat is the superintelligence itself,
as opposed to its human creators using it for destructive purposes?
Serious problems might arise even without assuming that superintel-
ligences have malice or even consciousness. Sufficiently powerful
optimization algorithms could pose a terrible risk. The problem with
computing machines is not that they won't do what they are programmed
to, but rather that they will do *exactly* what they are programmed to.
This is a problem because it can be hard to anticipate exactly what a
computer is programmed to do in a particular situation.

It is true that for comparatively simple programs—a word proc-
essor, for example, or a sorting algorithm—the programmer will precisely specify what the program should do in every situation. But as
we have discussed elsewhere, this is not how machine learning works.
Consider the seemingly easy task of distinguishing pictures of cats from
pictures of dogs. It is too difficult for anyone to enunciate precisely how
to perform this task, even though three-year-olds can solve the problem
easily and reliably. Instead, the way image classifiers work is that the
programmer specifies two things: a set of models to optimize over (say,
neural networks) and the objective function to be optimized (say, classification error). The actual computer program or model that results—
the image classifier that can distinguish cats from dogs—was not explicitly written by the programmer. Instead, it was "solved for" by

automatically optimizing the programmer's specified objective function over the class of models he specified. This methodology is very powerful, but sometimes it can lead to unanticipated behavior—like a navigation app sending drivers into a fire.

The fear is that as optimization technology becomes increasingly powerful—able to optimize over more and more complicated sets of models—it will become harder to anticipate the consequences of optimizing objective functions that we specify, even if we are careful.

CLOUDED VISION

To see this, let's consider a mundane story of machine learning gone awry. This hard-to-source story goes back to the earliest days of machine learning, in the 1960s. It has been used as a parable for illustrating the pitfalls of machine learning since at least the early 1990s. Here is a version from 1992:

> In the early days of the perceptron the army decided to train an artificial neural network to recognize tanks partly hidden behind trees in the woods. They took a number of pictures of a woods without tanks, and then pictures of the same woods with tanks clearly sticking out from behind trees. They then trained a net to discriminate the two classes of pictures. The results were impressive, and the army was even more impressed when it turned out that the net could generalize its knowledge to pictures from each set that had not been used in training the net.
>
> Just to make sure that the net had indeed learned to recognize partially hidden tanks, however, the researchers took some more pictures in the same woods and showed them to the trained net. They were shocked and depressed to find that with the new pictures the net totally failed to discriminate between

pictures of trees with partially concealed tanks behind them and just plain trees. The mystery was finally solved when someone noticed that the training pictures of the woods without tanks were taken on a cloudy day, whereas those with tanks were taken on a sunny day. The net had learned to recognize and generalize the difference between a woods with and without shadows!

The point of this story is that it can be hard to tell a computer what you want it to do when you are using machine learning. The way supervised learning works is that you give the algorithm lots of examples of the decisions you'd like it to make—in the parable, photographs together with their desired labels. The issue is that this can be an underspecified problem, even if it doesn't seem so. To a human being with knowledge of the context (the army would like a tool to be able to detect tanks), it is clear that between the two sets of images, the salient difference is that one set contains tanks and the other does not. But to the computer, *all* ways of distinguishing the two sets of images are equally good. If detecting shadows is easier and more accurate than detecting tanks, that is what it will do. With 1960s-era machine learning (and mostly with modern technology as well), this problem is not so hard to discover and to correct for. But the concern is that with more powerful technologies, the same basic pitfall could pose a much bigger problem.

Extrapolating a bit, we can imagine what might go wrong for similar reasons when we unleash an algorithm designed to optimize for a seemingly mundane task. Suppose we program a superpowerful optimization algorithm for a simple goal: mine as many bitcoins as possible within the next decade. Mining bitcoins requires solving a difficult computational problem, for which it is thought that there is no algorithm substantially better than brute-force search. One strategy that the algorithm could employ is to devote all of its computational power directly to these brute-force search problems and solve as many of them as possible within the next decade. This is what existing bitcoin miners do. But the algorithm's objective motivates it to find a better solution if one is available. With a bit of inspiration from science fiction, you can imagine dystopian solutions that would be improvements on the algorithm's narrowly defined objective function but which we didn't intend—including the forced reorientation of society's resources and even human civilization toward building bitcoin-mining rigs.

There are a couple of simple objections to these kinds of doomsday scenarios, but many of them can be dispatched with a little imagination. Perhaps the most obvious is "Why don't we just turn the computer off once we realize it is starting to exhibit these unintended behaviors?" But if the computer is turned off, it will have mined fewer bitcoins than if it had been left on. And remember, the computer is running a superpowerful optimization algorithm, so it is unlikely to miss this simple observation. So it should take steps to prevent anyone from turning it off—not because it has any particular instinct for self-preservation, but rather because turning it off would get in the way of optimizing its objective. (Some readers might be reminded here of the poignant death of the computer HAL in Stanley Kubrick's film *2001: A Space Odyssey*.)

A more robust solution would be to make sure that the objective the algorithm is optimizing has no undesirable side effects at all— that is, to fully align its objective function with our own. This is known as the value alignment problem: given an optimization algorithm, how do we set things up so that the optimization of its objective function results in outcomes that we actually want, *in all respects*? But as we have seen in more concretely grounded examples throughout this book, the value alignment problem is fiendishly difficult. Optimizing simple models for sensible objectives can lead to unanticipated violations of fairness and privacy, and to unanticipated feedback loops that lead to unintended collective behavior. If we can't even anticipate the effects of optimizing simple objective functions with the comparatively dumb learning algorithms that we have today, how much worse will things be if we use optimization engines that are substantially more intelligent than we are?

WHAT, ME WORRY?

The doomsday scenarios that one comes across when reading about intelligence explosions are dire, but necessarily fanciful and imprecise. Do they represent a real, concrete threat, or are they just material for science fiction novels? And even if the threat will eventually be real, are we in a position to productively think about it now? Some prominent AI experts think the answer to this last question is no. Andrew Ng, a former Stanford professor who went on to found the Google Brain group and served as chief scientist at Baidu, has said that worrying right now about the dangers of superintelligence is like worrying about overpopulation on Mars. In other words, it is something that could *conceivably* one day become a problem, but is distant enough from our current situation that we cannot productively think about it—at least compared to other things we might devote our valuable time to. The opposing view is that if we don't take the idea of an exponential superintelligence explosion seriously now, it might always seem like the problem is far in the future—until it is too late.

Ultimately, the crux of the question is whether we believe that the argument for a "fast takeoff"—a chain reaction that quickly builds on itself, resulting in increasingly intelligent machines, faster than we can keep pace with or even detect—is credible or not. So let's dive a bit deeper into this idea.

The basic supposition rests on the idea that all else being equal, a more powerful optimization algorithm will make more rapid progress in further developing machine learning technologies than a less powerful algorithm. It is hard to argue with this. But it's important to think about the *rates* at which this improvement happens—that is, what the learning curve is for designing increasingly powerful machine learning algorithms.

Many things in life exhibit diminishing marginal returns—the more you put in, the more you get out, but the less additional unit of output you get for each additional unit of input you put in. For example, human enjoyment of goods is like this: the first $10 million I give you improves your life much more dramatically than the next $10 million, and the first slice of chocolate cake you eat is more enjoyable than the sixth slice you eat in one sitting. Creative endeavors are generally like this too: the first hundred hours you put into writing a book chapter improves it more than the next hundred hours you spend revising. It might be that each additional hour spent revising the chapter makes it a little better, but it is not as if doubling the amount of time spent on the chapter will double its quality. Eventually your time will be better spent doing something else.

Other things don't exhibit diminishing marginal returns. Tasks demanding brute force are the clearest example. Assuming your well is full enough, doubling the amount of time you spend pumping water will double the amount of water you have. Here, returns to investment are linear: the amount of output you get is directly proportional to the amount of input. Other tasks actually exhibit increasing marginal returns. If you are reading *The Ethical Algorithm* on your Kindle, the marginal cost to produce your copy was essentially nothing: it just involved the copying and transmission of bits of information. But as we can attest, the effort involved in producing the *first* copy of the book was substantial. So whether we should expect a chain reaction leading to an exponential rate of improvement in learning algorithms or not should depend on whether we think that intelligence is more like eating chocolate cake, pumping water, or producing ebooks.

Suppose researching AI is more like pumping water: the rate at which we can make research progress in machine learning is exactly equal to the amount of machine intelligence we already have. If you write down a stylized model based on some simple differential equations, it turns out that this is the scenario in which we expect a "fast takeoff." If the return on research investment is linear, then the growth in intelligence is exponential. It is hard to even visualize this kind of growth; when plotted, it seems to stay constant until the very end, when it just shoots up. This is the scenario in which we might want to invest heavily in thinking about AI risk now—even though it appears that we are a long way off from superintelligence, it will keep looking like that until it is too late.

But suppose researching AI is more like writing an essay. In other words, it exhibits diminishing marginal returns. The more research we plow into it, the more intelligence we get, but at a slowing rate. It is less obvious what happens now, since as intelligence increases, the rate at which we produce new research increases as well—but the rate at which this new research actually manifests itself as increased intelligence is slowing. It's not too hard to convince yourself—again, playing with mathematical models of growth—that in this case, the rate of progress won't be anything close to exponential.

So should we expect exponential growth or not? This is closely related to the question of whether we expect to see diminishing returns on AI research. At the very least, it seems that an intelligence explosion is perhaps not as certain as I.J. Good made it out to be. And even if we do see exponential growth for a little while, we should remember that not all exponentials are equal. Doubling every day is an exponential rate of growth. But so is growing by 1 percent a year. And to paraphrase Ed Felten, a professor of computer science at Princeton, we should not simply deposit a few dollars in an interest-bearing savings account and then plan our retirement assuming that we will soon experience a "wealth explosion" that will suddenly make us unimaginably rich. Exponential growth can still seem slow on human time scales.

But even if an intelligence explosion is not certain, the fact that it remains a possibility, together with the potentially dire consequences it would entail, make methods for managing AI risk worth taking seriously as a topic of algorithmic research. After all, the core concern—that it is hard to anticipate the unintended side effects of optimizing seemingly sensible objectives—has been one of the driving forces behind every example of algorithmic misbehavior we have discussed in this book.

SOME CONCLUDING THOUGHTS

ONE CAUSE, MANY PROBLEMS

The last topic we covered in this book—the potential eventual risks of machine learning and AI to human safety, life, and even existence—might seem very different from the immediate and concrete risks we wrote more extensively about: loss of privacy, discrimination, gaming, and false discovery. For each of these narrower topics, we were able to discuss real-world examples in which actual algorithmic decision-making went wrong in some way, and then we explored specific algorithmic remedies. In Chapter 5, we had to speculate about what is currently science fiction.

But as we have pointed out, the life-threatening risks are just the logical extreme of the same basic problem that we grappled with in the other chapters—that blind, data-driven algorithmic optimization of a seemingly sensible objective can lead to unexpected and undesirable side effects. When we use a machine learning algorithm to optimize only for predictive accuracy, we shouldn't find it surprising when it produces a model that has wildly different false positive rates when applied to different demographic groups. We shouldn't find it surprising when it produces a model that encodes the identities of the individuals whose data was used for training, when it incentivizes people to misreport their data, or when it turns out to be gameable by data analysts seeking to make their research findings look more significant than they are. In all these cases, we are seeing different facets of the same problem. Optimization procedures—especially those optimizing over complicated domains—often lead to inscrutable outcomes that measure up exceptionally well when examined according to their objective function but generally will not satisfy constraints that weren't explicitly encoded in their design. And so if there are constraints that we want our algorithms to satisfy—because they encode some kind of ethical norm or, more dramatically, because the lack of such constraints poses an existential threat—we need to think precisely about what they are and how to embed them directly into the design of our algorithms.

ALGORITHMIC PROMISE

A solution that might seem attractive when one is initially bombarded with stories of algorithmic decision-making gone awry is to try to avoid algorithms altogether, at least when it comes to decisions of any importance. But while caution and prudence are warranted when deploying new technologies in consequential domains—and we are certainly far from fully understanding how machine learning interacts with issues such as fairness and others we have examined— avoiding algorithms is not a good long-term solution. This is for at least two reasons.

First, *all* decision-making—including that carried out by human beings—is ultimately algorithmic. The difference is that human decision-making is based on logic or behaviors that we struggle to precisely enunciate. If we humans had the ability to describe our own decision-making processes precisely enough, then we could in fact represent them as computer algorithms. So the choice is not whether to avoid using algorithms or not, but whether or not we should use *precisely specified* algorithms.

All things being equal, we should prefer being precise about what we are doing. Precision allows us, for example, to reason about counterfactuals: Would the outcome of your loan application have been different had your salary been higher by an additional $10,000 a year? Or if you were of a different race or gender? Human beings are good at coming up with self-reported explanations for their decisions, but as often as not, these explanations offer post hoc rationalizations instead of real insight. A lender who lets race sway a decision can still come up, after the fact, with a plausible, race-independent justification for denying an applicant a loan. We can only be sure about what decisions would have been made in different situations if we can pin down the decision-making process.

We must remember, of course, that all things are not equal: it is difficult to precisely specify good decision-making procedures and to gather rich, structured data, and the result is often that computer algorithms are designed to make decisions based on simpler and more restricted kinds of information than people have at their disposal. But this should be viewed as a challenge to overcome and a motivation for algorithmic research, rather than a blanket indictment of algorithms.

Second, machine learning is a powerful tool that has many extant and potential benefits. Technology companies such as Google and Facebook of course rely on products powered by machine learning for much of their revenue—but as these techniques grow in applicability, their scope and societal benefits grow as well. Rather than simply improving the click-through rate of targeted advertising, if learning procedures can improve the accuracy and efficacy of personalized medicine, they can potentially save lives. If they can predict creditworthiness from a broader set of indicators than just traditional measures such as income, savings, and credit card history, they can expand access to credit to a broader population. Realistic examples of this potential are endless. Although the risks of algorithmic decision-making are all too real, as we have shown throughout this book, avoiding those risks does not require giving up entirely on the benefits.

A related reaction is to assert that the solution to algorithmic misdeeds is more and better laws, regulations, and human oversight. Regulations and laws certainly have a crucial role to play—as we have emphasized throughout, the specification of what we want algorithms to do and not do for us should remain firmly in the human and societal arenas. But purely legal and regulatory approaches have a major problem: they don't scale. Any system that ultimately relies solely or primarily on human attention and oversight cannot possibly keep up with the volume and velocity of algorithmic decision-making. The result is that approaches that rely only on human oversight either entail largely giving up on algorithmic decision-making or will necessarily be outmatched by the scale of the problem and hence be insufficient. So while laws and regulations are important, we have argued in this book that the solution to the problems introduced by algorithmic decision-making should itself be in large part algorithmic.

This is not to say that we can have it all. One central lesson of this book is that additional constraints—like those imposed to correct ethical failures—won't come for free. There will always be trade-offs that we need to manage. Once we can precisely specify what we mean by "privacy" or "fairness," achieving these goals necessarily requires giving up on something else that we value—for example, raw predictive accuracy. It is the goal of algorithmic research not only to identify these constraints and embed them into our algorithms but also to quantify the extent of these trade-offs and to design algorithms that make them as mild as possible.

Deciding how best to manage these trade-offs, however, is not an algorithmic question but a social one best decided by stakeholders on a case-by-case basis. Is a gain in fairness—say, approximate equity of false positive rates across demographic groups—worth a particular drop in accuracy? What about an increase in the level of differential privacy? These questions don't have universal answers; the answers depend on the situation. When making life-or-death medical decisions, we might decide that accuracy is paramount. But when allocating spots in public high schools, we might feel fairness should take precedence. We might put an especially high premium on privacy over predictive accuracy when sensitive data about our social interactions is being used to better target advertisements. Ultimately, the decisions about where on the trade-off curve we want to live must be made by the people on the front lines. But defining the Pareto frontier itself—the space of optimal trade-offs—is a scientific problem.

IN THE BEGINNING...

A common criticism of technical work aimed at designing ethical algorithms is that it is akin to rearranging the deck chairs on the *Titanic*. While mathematicians debate the effects of tweaking the error statistics of machine learning algorithms, real injustice is being done by the very use of those algorithms in the first place. For example, if your family and neighbors are below the poverty line, then, statistically speaking, this really does make it less likely that you will be able to pay back a loan. An algorithm trained to be fair with respect to actual observed loan outcomes might be judged as correct in denying you a loan in part because of such considerations. But this is hardly fair in the grand scheme of things, because—to borrow a phrase from discrimination law expert Deborah Hellman—it compounds injustice.

What might appear fair from a myopic point of view is seen to be unfair when one takes into account the societal context: a lending algorithm designed like this would be part of a larger system that further punishes people for being poor, resulting in a feedback loop.

More generally, by myopically focusing on quantitative properties of an algorithm in a single, static context, we are ignoring aspects of the problem that are vital to understanding fairness: the downstream and upstream effects of the decisions that we make. After all, when we use algorithms not just to make predictions but to make decisions, they are changing the world in which they operate, and we need to take into account such dynamic effects in order to talk sensibly about something like "fairness."

While there is a great deal of truth to this view, it should not be an indictment of ethical algorithms generally. Instead, it highlights the complexity of the problems and the embryonic state of algorithmic research in these areas. Many aspects of the current mathematical literature on algorithmic fairness and privacy can seem naive on the surface because of the simplicity of the abstractions they propose and their limited purview. But it is inherent in the rigorous approach that we propose—precisely specifying one's goals and then designing algorithms that achieve them—that we must start as simply as possible. Precisely specifying one's goals is *hard work*. The more complex the issue at hand, the easier it is to hide from it in the fog of ambiguity. But doing so doesn't solve the ethical problems that we set out to fix; it only obscures them. Of course, the solution is not to replace nuanced but vague human decision-making with the rigor and precision of an oversimplified model. But by starting to formalize our goals in simple—even oversimplified—scenarios, we can gradually work our way up to the real and complex problems we want to address.

The result is that, at least for a while, the critics of the algorithmic approach may often be right. There are many consequential domains where algorithmic tools are still too naive and primitive to be fully trusted with decision-making. This is because to model the forest, we need to start with the trees. This book offers a snapshot of exciting strands of research aimed at developing ethical algorithms, many of which are still in their very earliest days. What we advocate is scientific methodology, driven by precise definitions, and not any particular existing technology.

At its best, this methodology can bear valuable fruit. Differential privacy is a case in point. Over the past fifteen years, differential privacy has transitioned from a purely academic curiosity among theoretical computer scientists to a maturing technology trusted with protecting the statistical disclosures for the 2020 US Census and deployed at a large scale on iPhones and Google Chrome web browsers. Differential privacy offered a change from its predecessors in its precise definition of what it was supposed to guarantee, and this greatly aided its development. Of course, differential privacy does not promise us everything that we might mean when we use the English word *privacy*; no single definition could. But the precision of its definition aids us in delineating exactly when its promises are, and are not, what we want. The privacy issue is far from being resolved, but at least it is in a state where we can speak rigorously about which kinds of problems can and cannot be solved by current technology.

If we can achieve the same thing for the other ethical concerns we study in this book, we will be in good stead. It cannot and should not be rushed: despite breathless journalism, progress will come in fits and starts, and will take years. But ethical algorithms hold great promise. We are just at the beginning of a fascinating and important journey.

ACKNOWLEDGMENTS

Every aspect of this book—the development of the underlying research, the themes and narratives, the writing and the production—has benefited greatly from the help of a number of colleagues, friends, professionals, and institutions. We'd like to take some space to give our warm thanks to them.

On the research side, and starting closest to home, our ideas and work on the topics in these pages has been deeply influenced by a close-knit group of current and former Penn graduate students, postdocs, and visitors that are like a second family to us. They have helped shape our thoughts on everything from the right mathematical models for ethical algorithms, the design and analysis of those algorithms, and the higher-level considerations of how they do or don't address real societal problems. This merry band of friends includes Rachel Cummings, Jinshou Dong, Hadi Elzayn, Hoda Heidari, Justin Hsu, Zhiyi Huang, Shahin Jabbari, Matthew Joseph, Chris Jung, Jieming Mao, Jamie Morgenstern, Seth Neel, Ryan Rogers, Zachary Schutzman, Saeed Sharifi, Bo Waggoner, Steven Wu, Grigory Yaroslavtzev, and Juba Ziani. Thanks, Gang!

We also give hearty thanks to external research colleagues whom we have worked closely with or who have deeply influenced our ideas, including Cynthia Dwork, Vitaly Feldman, Moritz Hardt, Mike Jordan, Jon Kleinberg, Katrina Liggett, Kobbi Nissim, Mallesh Pai, Toni Pitassi, Omer Reingold, Tim Roughgarden, Sebastian

Seung, Adam Smith, Jonathan Ullman, Salil Vadhan, Jenn Wortman Vaughan, and Duncan Watts.

A number of Penn faculty members, both in our own department and in diverse areas such as law, criminology, and economics, have collaborated with us and helped broaden our perspective on the relationships between algorithms and society, including Tom Baker, Richard Berk, Cary Coglianese, Ezekiel Dixon-Román, Andreas Haeberlen, Sampath Kannan, Sanjeev Khanna, Annie Liang, Ani Nenkova, Benjamin Pierce, Rakesh Vohra, and Christopher Yoo.

Penn leadership, both current and past, has provided a tremendously supportive, flexible, and productive environment for us as researchers and educators. These days virtually every major research university makes grand claims to interdisciplinarity, but Penn is the real deal. For that we give warm thanks to Eduardo Glandt, Amy Gutmann, Vijay Kumar, Vincent Price, and Wendell Pritchett. We are particularly grateful to Fred and Robin Warren, founders and benefactors of Penn's Warren Center for Network and Data Sciences, for helping to create the remarkable intellectual melting pot that allowed this book to develop. Many thanks to Lily Hoot of the Warren Center for her unflagging professionalism and organizational help. We also are grateful to Raj and Neera Singh, founders and benefactors of Penn's Networked and Social Systems Engineering (NETS) Program, in which we developed much of the narrative expressed in these pages.

A number of important professional and personal figures have played major roles in the writing and production of this book. We'd like to warmly thank Sarah Humphreville, our editor at Oxford University Press, for her careful and helpful readings and suggestions, which have always been on the mark. Many thanks to Joellyn Ausanka of OUP for her expert handling of the production process. We thank Eric Henney of Basic Books for his early encouragement of this project and our agent, Jim Levine, for helping us navigate the Byzantine world of trade publishing. We are grateful for comments on early drafts from Thomas Kearns, Yuriy Nevmyaka, Alvin Roth, and Ben Roth.

Even though both of us are many years past our doctoral studies, we were deeply and forever influenced by the outstanding guidance of our own dissertation advisors, Avrim Blum (AR) and Les Valiant (MK). A transition every graduate student in our research areas must make is that from merely solving given problems to choosing one's own projects and to developing research "taste"—deciding what's important to work on from a broader, nontechnical perspective. Both of us feel we can trace the ideas in this book to the aesthetic mentorship of our advisors.

Finally, as was declared in the earliest pages, this book is dedicated to our families—Kim, Kate, and Gray (MK), and Cathy, Eli, and Zelda (AR). Both the book and the underlying ideas and research would have been inconceivable without the warm, loving, and fun environments we enjoy away from work. Thanks and much love to you all!

REFERENCES AND
FURTHER READING

CHAPTER 1: ALGORITHMIC PRIVACY:
FROM ANONYMITY TO NOISE

References

An extended discussion of successful "de-anonymization" attacks, including the Massachusetts GIC and Netflix cases, can be found in "Broken Promises of Privacy: Responding to the Surprising Failure of Anonymization" by Paul Ohm, which appeared in the *UCLA Law Review* 57 (2010). Details on the Netflix attack are described in "Robust De-anonymization of Large Sparse Datasets" by Arvind Narayanan and Vitaly Shmatikov, which was published in the *IEEE Symposium on Security and Privacy* (IEEE, 2008). Details of the original Genome-Wide Association Study attack can be found in "Resolving Individuals Contributing Trace Amounts of DNA to Highly Complex Mixtures Using High-Density SNP Genotyping Microarrays" by Nils Homer, Szabolcs Szelinger, Margot Redman, David Duggan, Waibhav Tembe, Jill Muehling, John V. Pearson, Dietrich A. Stephan, Stanley F. Nelson, and David W. Craig, which appeared in *PLOS Genetics* (2008). The notion of k-anonymity was first proposed by Latanya Sweeney in "k-anonymity: A model for protecting privacy," which appeared in the *International Journal of Uncertainty, Fuzziness, and Knowledge-Based Systems* (World Scientific, 2002).

The notion of differential privacy was first introduced in "Calibrating Noise to Sensitivity in Private Data Analysis" by Cynthia Dwork, Frank McSherry, Kobbi Nissim, and Adam Smith, which appeared in *Theory of Cryptography: Third Theory of Cryptography Conference* (Springer, 2006). The Bayesian interpretation of differential privacy is due to "On the 'Semantics' of Differential Privacy: A Bayesian Formulation" by Shiva Prasad Kasiviswanathan and Adam Smith, which appeared in the *Journal of Privacy and Confidentiality* 6 (2014). Randomized response was introduced in "Randomized Response: A Survey Technique for Eliminating Evasive Answer Bias" by Stanley L. Warner, published in the *Journal of the American Statistical Association* 60 (1965): 309.

The study showing that a user's pattern of Facebook likes correlates to various sensitive personal characteristics appears in "Private Traits and Attributes Are Predictable from Digital Records of Human Behavior" by Michal Kosinski, David Stillwell, and Thore Graepel, published in the *Proceedings of the National Academy of Sciences* 110 (2013).

Further Reading

An excellent general-audience overview of modern data collection, security, and privacy concerns can be found in the book *Data and Goliath: The Hidden Battles to Collect Your Data and Control Your World* by Bruce Schneier (W. W. Norton, 2015).

A comprehensive technical introduction to the definitions, methods, and algorithms of differential privacy can be found in *The Algorithmic Foundations of Differential Privacy* by Cynthia Dwork and Aaron Roth (NOW Publishers, 2014).

CHAPTER 2: ALGORITHMIC FAIRNESS: FROM PARITY TO PARETO

References

The research on gender bias in word embeddings was first described in "Man Is to Computer Programmer as Woman Is to Homemaker? Debiasing Word Embeddings," by Tolga Bolukbasi, Kai-Wei Chang, James Zou, Venkatesh Saligrama, and Adam Kalai, which was presented at the 30th Conference on Neural Information Processing Systems, 2016. Similar findings were reported in "Semantics Derived Automatically from Language Corpora Contain Human-Like Biases" by Aylin Caliskan, Joanna J. Bryson, and Arvind Narayanan, *Science* 356, no. 6344 (2017).

A foundational paper outlining many of the basic issues surrounding fairness in machine learning, including that "forbidding" the use of certain inputs is doomed to failure, is "Fairness Through Awareness" by Cynthia Dwork, Moritz Hardt, Toniann Pitassi, Omer Reingold, and Richard Zemel, which appeared in *Proceedings of the 3rd Innovations in Theoretical Computer Science Conference* (ACM, 2012).

Research first demonstrating the impossibility of simultaneously satisfying multiple fairness definitions is described in "Fair Prediction with Disparate Impact: A Study of Bias in Recidivism Prediction Instruments" by Alexandra Chouldechova, which was presented at the Workshop on Fairness, Accountability and Transparency in Machine Learning, 2016; and in "Inherent Trade-Offs in the Fair Determination of Risk Scores" by Jon Kleinberg, Sendhil Mullainathan, and Manish Raghavan, which was presented at the 8th Innovations in Theoretical Computer Science Conference, 2017. Equalizing false positive rates and/or false negative rates as a fairness constraint was proposed in "Equality of Opportunity in Supervised Learning," by Moritz Hardt, Eric Prize, and Nati Srebro, which was presented at *Advances in Neural Information Processing*, 2016.

The fairness-accuracy Pareto curves and the notion of fairness gerrymandering are from the papers "Preventing Fairness Gerrymandering: Auditing and Learning for Subgroup Fairness," presented at the International Conference on Machine Learning, 2018, and "An Empirical Study of Subgroup Fairness for Machine Learning," presented at the ACM Conference on Fairness, Accountability and Transparency, 2019, both by Michael Kearns, Seth Neel, Aaron Roth, and Zhiwei Steven Wu.

Further Reading

Recently a number of good general-audience books on algorithmic discrimination and its consequences have been published. These include *Weapons of Math Destruction: How Big Data Increases Inequality and Threatens Democracy* by Cathy O'Neil (Crown, 2016), *Algorithms of Oppression: How Search Engines Reinforce Racism* by Safiya Umoja Noble (New York University Press, 2018), and *Automating Inequality: How High-Tech Tools Profile, Police, and Punish the Poor* by Virginia Eubanks (St. Martin's Press, 2017).

For a more technical overview of recent research in algorithmic fairness with many references to source materials, see the article "The Frontiers of Fairness in Machine Learning" by Alexandra Chouldechova and Aaron Roth, 2018. A textbook treatment by Solon Barocas, Moritz Hardt, and Arvind Narayanan is forthcoming from MIT Press and can be found at https://fairmlbook.org.

CHAPTER 3: GAMES PEOPLE PLAY (WITH ALGORITHMS)

References

Many of the observations and results regarding the commuting game and the Maxwell solution are summarized in *Selfish Routing and the Price of Anarchy* by Tim Roughgarden (MIT Press, 2005). The ideas behind Maxwell 2.0 and their connection to differential privacy are described in "Mechanism Design in Large Games" by Michael Kearns, Mallesh M. Pai, Aaron Roth, and Jonathan Ullman, which appeared in *Proceedings of the 2014 Conference on Innovations in Theoretical Computer Science* (ACM, 2014).

The Gale-Shapley algorithm was first described in "College Admissions and the Stability of Marriage" by David Gale and Lloyd S. Shapley, in *American*

Mathematical Monthly 69 (1962). An overview of the work on applying matching algorithms to kidney exchange is given in *Who Gets What—and Why* by Alvin E. Roth (Houghton Mifflin Harcourt, 2015).

The application of algorithmic self-play to the development of a backgammon program was described in "Temporal Difference Learning and TD-Gammon" by Gerald Tesauro, which appeared in *Communications of the ACM* 38 (1995). Generative adversarial networks were first described in "Generative Adversarial Nets" by Ian J. Goodfellow, Jean Pouget-Abadie, Mehdi Mirza, Bing Xu, David Warde-Farley, Sherjil Ozair, Aaron Courville, and Yoshua Bengio, which appeared in *Neural Information Processing: 21st International Conference* (Springer, 2014).

The framing of private synthetic data generation as computation of an equilibrium in a game was first given in "Differential Privacy for the Analyst via Private Equilibrium Computation," by Justin Hsu, Aaron Roth, and Jonathan Ullman, presented at the 45th Annual ACM Symposium on Theory of Computing (2013). Work on using differentially private GANs to construct synthetic medical data was done in "Privacy-Preserving Generative Deep Neural Networks Support Clinical Data Sharing" by Brett K. Beaulieu-Jones, Zhiwei Steven Wu, Chris Williams, Ran Lee, Sanjeev P. Bhavnani, James Brian Byrd, and Casey S. Greene, available on bioArxiv.

Further Reading

Good technical treatments of many of the topics discussed in this chapter can be found in the collection of articles *Algorithmic Game Theory* edited by Noam Nisan, Tim Roughgarden, Eva Tardos, and Vijay V. Vazirani (Cambridge University Press, 2007), and *Twenty Lectures on Algorithmic Game Theory* by Tim Roughgarden (Cambridge University Press, 2016).

CHAPTER 4: LOST IN THE GARDEN: LED ASTRAY BY DATA

References

John Ioannidis was amongst the first to model the statistical crisis in science, with his influential paper "Why Most Published Research Findings Are False," which appeared in *PLoS Medicine* (2005). The term "garden of the forking paths" in the context of this chapter comes from an essay by Andrew Gelman and Erik Loken, called "The Statistical Crisis in Science," which appears in *American Scientist* 102, no. 6 (2014). This essay also includes a number of illustrative examples of the phenomenon. We quote "The Preregistration Revolution" by Brian A. Nosek, Charles R. Ebersole, Alexander DeHaven, and David Mellor, which appeared in *Proceedings of the National Academy of Sciences of the United States of America* 115, no. 11 (2018).

The leaderboard algorithm for machine learning competitions we describe is due to a paper of Avrim Blum and Moritz Hardt, called "The Ladder: A Reliable Leaderboard for Machine Learning Competitions," presented at the International Conference on Machine Learning (2015). The connection between differential privacy and preventing overfitting was made by Cynthia Dwork, Vitaly Feldman, Moritz Hardt, Toniann Pitassi, Omer Reingold, and Aaron Roth, in a paper called "The Reusable Holdout: Preserving Validity in Adaptive Data Analysis," which appeared in *Science* 349, no. 6248 (2015). The connection between "summarizability" and preventing overfitting was made by Cynthia Dwork, Vitaly Feldman, Moritz Hardt, Toniann Pitassi, Omer Reingold, and Aaron Roth in a paper called "Generalization in Adaptive Data Analysis and Holdout Reuse," which appeared in *Neural Information Processing Systems: 22nd International Conference* (Springer, 2015).

Further Reading

For a thorough overview of the mathematics behind the algorithmic techniques discussed in this chapter, we refer the reader to a set of lecture notes for a class on "Adaptive Data Analysis," taught by Aaron Roth and Adam Smith, available at http://www.adaptivedataanalysis.com.

CHAPTER 5: RISKY BUSINESS: INTERPRETABILITY, MORALITY, AND THE SINGULARITY

References

Research on the interpretability of neural networks and "cat neurons" appeared in "Building High-Level Features Using Large Scale Unsupervised Learning" by Quoc V. Le, Marc'Aurelio Ranzato, Rajat Monga, Matthieu Devin, Kai Chen, Greg S. Corrado, Jeff Dean, and Andrew Y. Ng, which appeared in *Proceedings of the 29th International Conference on Machine Learning* (IMLS, 2012).

The work and thoughts of Michael Sandel on morality and ethics is overviewed in his books *What Money Can't Buy: The Moral Limits of Markets* (Farrar, Straus, and Giroux, 2013), and *Justice: What's the Right Thing to Do* (Farrar, Straus, and Giroux, 2009).

INDEX